GEOMETRIC TRANSFORMATIONS I

NEW MATHEMATICAL LIBRARY

published by
The Mathematical Association of America

The New Mathematical Library (NML) was begun in 1961 by the School Mathematics Study Group to make available to high school students short expository books on various topics not usually covered in the high school syllabus. In a decade the NML matured into a steadily growing series of some twenty titles of interest not only to the originally intended audience, but to college students and teachers at all levels. Previously published by Random House and L. W. Singer, the NML became a publication series of the Mathematical Association of America (MAA) in 1975. Under the auspices of the MAA the NML will continue to grow and will remain dedicated to its original and expanded purposes.

GEOMETRIC TRANSFORMATIONS I

by

I. M. Yaglom

translated from the Russian by

Allen Shields
University of Michigan

8

THE MATHEMATICAL ASSOCIATION
OF AMERICA

Ninth Printing

 Published in New York by Random House, Inc., and simultaneously in Toronto, Canada, by Random House of Canada, Limited.
School Edition Published by The L. W. Singer Company.

Library of Congress Catalog Card Number: 62-18330

Complete Set ISBN-0-88385-600-X
Vol. 8 0-88385-608-5

Manufactured in the United States of America

Note to the Reader

This book is one of a series written by professional mathematicians in order to make some important mathematical ideas interesting and understandable to a large audience of high school students and laymen. Most of the volumes in the *New Mathematical Library* cover topics not usually included in the high school curriculum; they vary in difficulty, and, even within a single book, some parts require a greater degree of concentration than others. Thus, while the reader needs little technical knowledge to understand most of these books, he will have to make an intellectual effort.

If the reader has so far encountered mathematics only in classroom work, he should keep in mind that a book on mathematics cannot be read quickly. Nor must he expect to understand all parts of the book on first reading. He should feel free to skip complicated parts and return to them later; often an argument will be clarified by a subsequent remark. On the other hand, sections containing thoroughly familiar material may be read very quickly.

The best way to learn mathematics is to *do* mathematics, and each book includes problems, some of which may require considerable thought. The reader is urged to acquire the habit of reading with paper and pencil in hand; in this way mathematics will become increasingly meaningful to him.

For the authors and editors this is a new venture. They wish to acknowledge the generous help given them by the many high school teachers and students who assisted in the preparation of these monographs. The editors are interested in reactions to the books in this series and hope that readers will write to: Editorial Committee of the NML series, New York University, The Courant Institute of Mathematical Sciences, 251 Mercer Street, New York, N. Y. 10012.

The Editors

NEW MATHEMATICAL LIBRARY

1 Numbers: Rational and Irrational *by Ivan Niven*
2 What is Calculus About? *by W. W. Sawyer*
3 An Introduction to Inequalities *by E. F. Beckenbach and R. Bellman*
4 Geometric Inequalities *by N. D. Kazarinoff*
5 The Contest Problem Book I Annual High School Mathematics Examinations 1950–1960. Compiled and with solutions *by Charles T. Salkind*
6 The Lore of Large Numbers, *by P. J. Davis*
7 Uses of Infinity *by Leo Zippin*
8 Geometric Transformations I *by I. M. Yaglom, translated by A. Shields*
9 Continued Fractions *by Carl D. Olds*
10 Graphs and Their Uses *by Oystein Ore*
11, 12 Hungarian Problem Books I and II, Based on the Eötvös Competitions 1894–1905 and 1906–1928, *translated by E. Rapaport*
13 Episodes from the Early History of Mathematics *by A. Aaboe*
14 Groups and Their Graphs *by I. Grossman and W. Magnus*
15 The Mathematics of Choice *by Ivan Niven*
16 From Pythagoras to Einstein *by K. O. Friedrichs*
17 The Contest Problem Book II Annual High School Mathematics Examinations 1961–1965. Compiled and with solutions *by Charles T. Salkind*
18 First Concepts of Topology *by W. G. Chinn and N. E. Steenrod*
19 Geometry Revisited *by H. S. M. Coxeter and S. L. Greitzer*
20 Invitation to Number Theory *by Oystein Ore*
21 Geometric Transformations II *by I. M. Yaglom, translated by A. Shields*
22 Elementary Cryptanalysis—A Mathematical Approach *by A. Sinkov*
23 Ingenuity in Mathematics *by Ross Honsberger*
24 Geometric Transformations III *by I. M. Yaglom, translated by A. Shenitzer*
25 The Contest Problem Book III Annual High School Mathematics Examinations 1966–1972. Compiled and with solutions *by C. T. Salkind and J. M. Earl*
26 Mathematical Methods in Science *by George Pólya*
27 International Mathematical Olympiads—1959–1977. Compiled and with solutions *by S. L. Greitzer*
28 The Mathematics of Games and Gambling *by Edward W. Packel*
29 The Contest Problem Book IV Annual High School Mathematics Examinations 1973–1982. Compiled and with solutions *by R. A. Artino, A. M. Gaglione and N. Shell*
30 The Role of Mathematics in Science *by M. M. Schiffer and L. Bowden*
31 International Mathematical Olympiads 1978–1985 and forty supplementary problems. Compiled and with solutions by *Murray S. Klamkin.*
32 Riddles of the Sphinx *by Martin Gardner*
33 U.S.A. Mathematical Olympiads 1972–1986. Compiled and with solutions by *Murray S. Klamkin.*
34 Graphs and Their Uses, *by Oystein Ore*. Revised and updated *by Robin J. Wilson.*

Other titles in preparation.

CONTENTS

GEOMETRIC TRANSFORMATIONS I

Translator's Preface

The present volume is Part I of *Geometric Transformations* by I. M. Yaglom. The Russian original appeared in three parts; Parts I and II were published in 1955 in one volume of 280 pages. Part III was published in 1956 as a separate volume of 611 pages. In the English translation Parts I and II are published as two separate volumes: NML 8 and NML21. The first chapter of Part III, on projective and some non-Euclidean geometry, was translated into English and published in 1973 as NML vol. 24; the balance of Part III, on inversions, has not so far been published in English.

In this translation most references to Part III were eliminated, and Yaglom's "Foreword" and "On the Use of This Book" appear, in greatly abbreviated form, under the heading "From the Author's Preface".

This book is not a text in plane geometry. On the contrary, the author assumes that the reader is already familiar with the subject. Most of the material could be read by a bright high school student who has had a term of plane geometry. However, he would have to work; this book, like all good mathematics books, makes considerable demands on the reader.

The book deals with the fundamental transformations of plane geometry, that is, with distance-preserving transformations (translations, rotations, reflections) and thus introduces the reader simply and directly to some important group theoretic concepts.

The relatively short basic text is supplemented by 47 rather difficult problems. The author's concise way of stating these should not discourage the reader; for example, he may find, when he makes a diagram of the given data, that the number of solutions of a given problem depends on the relative lengths of certain distances or on the relative positions of certain given figures. He will be forced to discover for himself the conditions under which a given problem has a unique solution. In the second half of this book, the problems are solved in detail and a discussion of the conditions under which there is no solution, or one solution, or several solutions is included.

The reader should also be aware that the notation used in this book may be somewhat different from the one he is used to. For example, if two lines l and m intersect in a point O, the angle between them is often referred to as $\sphericalangle lOm$; or if A and B are two points, then "the line AB" denotes the line through A and B, while "the line segment AB" denotes the finite segment from A to B.

The footnotes preceded by the usual symbol † were taken over from the Russian version of this book while those preceded by the symbol T have been added in this translation.

I wish to thank Professor Yaglom for his valuable assistance in preparing the American edition of his book. He read the manuscript of the translation and made a number of suggestions. He has expanded and clarified certain passages in the original, and has added several problems. In particular, Problems 4, 14, 24, 42, 43, and 44 in this volume were not present in the original version while Problems 22 and 23 of the Russian original do not appear in the American edition. In the translation of the next part of Yaglom's book, the problem numbers of the American edition do not correspond to those of the Russian edition. I therefore call to the reader's attention that all references in this volume to problems in the sequel carry the problem numbers of the Russian version. However, NML 21 includes a table relating the problem numbers of the Russian version to those in the translation (see p. viii of NML 21).

The translator calls the reader's attention to footnote † on p. 20, which explains an unorthodox use of terminology in this book.

Project for their advice and assistance. Professor H. S. M. Coxeter was particularly helpful with the terminology. Especial thanks are due to Dr. Anneli Lax, the technical editor of the project, for her invaluable assistance, her patience and her tact, and to her assistants Carolyn Stone and Arlys Stritzel.

Allen Shields

From the Author's Preface

This work, consisting of three parts, is devoted to elementary geometry. A vast amount of material has been accumulated in elementary geometry, especially in the nineteenth century. Many beautiful and unexpected theorems were proved about circles, triangles, polygons, etc. Within elementary geometry whole separate "sciences" arose, such as the geometry of the triangle or the geometry of the tetrahedron, having their own, extensive, subject matter, their own problems, and their own methods of solving these problems.

The task of the present work is not to acquaint the reader with a series of theorems that are new to him. It seems to us that what has been said above does not, by itself, justify the appearance of a special monograph devoted to elementary geometry, because most of the theorems of elementary geometry that go beyond the limits of a high school course are merely curiosities that have no special use and lie outside the mainstream of mathematical development. However, in addition to concrete theorems, elementary geometry contains two important general ideas that form the basis of all further development in geometry, and whose importance extends far beyond these broad limits. We have in mind the deductive method and the axiomatic foundation of geometry on the one hand, and geometric transformations and the group-theoretic foundation of geometry on the other. These ideas have been very fruitful; the development of each leads to non-Euclidean geometry. The description of one of these ideas, the idea of the group-theoretic foundation of geometry, is the basic task of this work. . . .

Let us say a few more words about the character of the book. It is intended for a fairly wide class of readers; in such cases it is always necessary to sacrifice the interests of some readers for those of others. The author has sacrificed the interests of the well prepared reader, and has striven for simplicity and clearness rather than for rigor and for logical exactness. Thus, for example, in this book we do not define the general concept of a geometric transformation, since defining terms that are intuitively clear always causes difficulties for inexperienced readers. For the same reason it was necessary to refrain from using directed angles and to postpone to the second chapter the introduction of directed segments, in spite of the disadvantage that certain arguments in the basic text and in the solutions of the problems must, strictly speaking, be considered incomplete (see, for example, the proof on page 50). It seemed to us that in all these cases the well prepared reader could complete the reasoning for himself, and that the lack of rigor would not disturb the less well prepared reader. . . .

The same considerations played a considerable role in the choice of terminology. The author became convinced from his own experience as a student that the presence of a large number of unfamiliar terms greatly increases the difficulty of a book, and therefore he has attempted to practice the greatest economy in this respect. In certain cases this has led him to avoid certain terms that would have been convenient, thus sacrificing the interests of the well prepared reader. . . .

The problems provide an opportunity for the reader to see how well he has mastered the theoretical material. He need not solve all the problems in order, but is urged to solve at least one (preferably several) from each group of problems; the book is constructed so that, by proceeding in this manner, the reader will not lose any essential part of the content. After solving (or trying to solve) a problem, he should study the solution given in the back of the book.

The formulation of the problems is not, as a rule, connected with the text of the book; the solutions, on the other hand, use the basic material and apply the transformations to elementary geometry. Special attention is paid to methods rather than to results; thus a particular exercise may appear in several places because the comparison of different methods of solving a problem is always instructive.

There are many problems in construction. In solving these we are not interested in the "simplest" (in some sense) construction—instead the author takes the point of view that these problems present mainly a logical interest and does not concern himself with actually carrying out the construction.

No mention is made of three-dimensional propositions; this restriction does not seriously affect the main ideas of the book. While a section of problems in solid geometry might have added interest, the problems in this book are illustrative and not at all an end in themselves.

The manuscript of the book was prepared by the author at the Orekhovo-Zuevo Pedagogical Institute . . . in connection with the author's work in the geometry section of the seminar in secondary school mathematics at Moscow State University.

I. M. Yaglom

INTRODUCTION

What is Geometry?

On the first page of the high school geometry text by A. P. Kiselyov,[T] immediately after the definitions of *point*, *line*, *surface*, *body*, and the statement "a collection of points, lines, surfaces or bodies, placed in space in the usual manner, is called a geometric figure", the following definition of geometry is given: "*Geometry is the science that studies the properties of geometric figures.*" Thus one has the impression that the question posed in the title to this introduction has already been answered in the high school geometry texts, and that it is not necessary to concern oneself with it further.

But this impression of the simple nature of the problem is mistaken. Kiselyov's definition cannot be called false; however, it is somewhat incomplete. The word "property" has a very general character, and by no means all properties of figures are studied in geometry. Thus, for example, it is of no importance whatever in geometry whether a triangle is drawn on white paper or on the blackboard; the color of the triangle is not a subject of study in geometry. It is true, one might answer, that geometry studies ***properties of geometric figures*** in the sense of the definition above, and that color is a property of the paper on which the figure is drawn, and is not a property of the figure itself. However, this answer may still leave a certain feeling of dissatisfaction; in order to carry greater conviction one would like to be able to quote a precise "mathematical" definition of exactly which properties of figures are

[T] This is the leading textbook of plane geometry in the Soviet Union.

studied in geometry, and such a definition is lacking. This feeling of dissatisfaction grows when one attempts to explain why it is that, in geometry, one studies the distance from a vertex of a triangle drawn on the board to certain lines, for example, to the opposite side of the triangle, and not to other lines, for example, to the edge of the board. Such an explanation can hardly be given purely on the basis of the definition above.

Before continuing with the presentation we should note that the school textbook cannot be reproached for the incompleteness of its definition. Kiselyov's definition is, perhaps, the only one that can be given at the first stage in the study of geometry. It is enough to say that the history of geometry begins more than 4000 years ago, and the first scientific definition of geometry, the description of which is one of the main aims of this book, was given only about 80 years ago (in 1872) by the German mathematician F. Klein. It required the creation of non-Euclidean geometry by Lobachevsky before mathematicians clearly recognized the need for an exact definition of the subject matter of geometry; only after this did it become clear that the intuitive concept of "geometric figures", which presupposed that there could not be several "geometries", could not provide a sufficient foundation for the extensive structure of the science of geometry.†

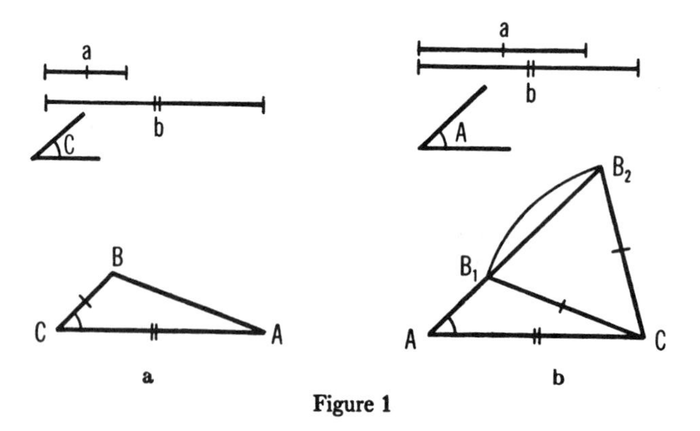

Figure 1

Let us now turn to the clarification of exactly which properties of geometric figures are studied in geometry. We have seen that geometry does not study all properties of figures, but only some of them; before having a precise description of those properties that belong to geometry

† Although non-Euclidean geometry provided the impetus that led to the precise definition of geometry, this definition itself can be fully explained to people who know nothing of the geometry of Lobachevsky.

we can only say that geometry studies "geometric properties" of figures. This addition to Kiselyov's definition does not of itself complete the definition; the question has now become, what are "geometric properties"? and we can answer only that they are "those properties that are studied in geometry". Thus we have gone around in a circle; we defined geometry as the science that studies geometric properties of figures, and geometric properties as being those properties studied in geometry. In order to break this circle we must define "geometric property" without using the word "geometry".

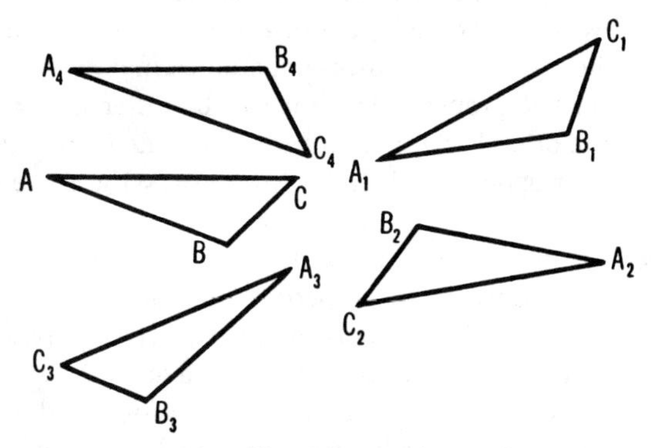

Figure 2

To study the question of what are "geometric properties" of figures, let us recall the following well known proposition: *The problem of constructing a triangle, given two sides a, b, and the included angle C, has only one solution* (Figure 1a).† On second thought, the last phrase may seem to be incorrect; there is really not just one triangle with the given sides a, b, and the included angle C, but there are infinitely many (Figure 2), so that our problem has not just one solution, but infinitely many. What then does the assertion, that there is just one solution, mean?

The assertion that from two sides a, b, and the included angle C *only one* triangle can be constructed clearly means that all triangles having the given sides a, b, and the included angle C are congruent to one another. Therefore it would be more accurate to say that from two sides and the included angle one can construct infinitely many triangles, but they are all congruent to one another. Thus in geometry when one says that there exists a unique triangle having the given sides a, b, and the included angle C, then triangles that differ only in their positions are not

† In contrast to this, the problem of constructing a triangle given the sides a, b, and the angle A opposite one of the given sides can have two solutions (Figure 1b).

considered to be different. And since we defined geometry as the science that studied "geometric properties" of figures, then clearly only figures that have exactly the same geometric properties will be indistinguishable from one another. Thus congruent figures will have exactly the same geometric properties; conversely, figures that are not congruent must have different geometric properties, for otherwise they would be indistinguishable.

Thus we have come to the required definition of geometric properties of figures: *Geometric properties of figures are those properties that are common to all congruent figures.* Now we can give a precise answer to the question of why, for example, the distance from one of the vertices of a triangle to the edge of the board is not studied in geometry: This distance is not a geometric property, since it can be different for congruent triangles. On the other hand, the altitude of a triangle is a geometric property, since corresponding altitudes are always the same for congruent figures.

Now we are much closer to the definition of geometry. We know that geometry studies "geometric properties" of figures, that is, those properties that are the same for congruent figures. It only remains for us to answer the question: "What are congruent figures?"

This last question may disappoint the reader, and may create the impression that thus far we have not achieved anything; we have simply changed one problem into another one, just as difficult. However, this is really not the case; the question of when two figures are congruent is not at all difficult, and Kiselyov's text gives a completely satisfactory answer to it. According to Kiselyov, "*Two geometric figures are said to be congruent if one figure, by being moved in space, can be made to coincide with the second figure so that the two figures coincide in all their parts.*" In other words, congruent figures are those that can be made to coincide by means of a motion; therefore, geometric properties of figures, that is, properties common to all congruent figures, are those properties that are not changed by moving the figures.

Thus we finally come to the following definition of geometry: *Geometry is the science that studies those properties of geometric figures that are not changed by motions of the figures.* For the present we shall stop with this definition; there is still room for further development, but we shall have more to say of this later on.

A nagging critic may not even be satisfied with this definition and may still demand that we define what is meant by a motion. This can be answered in the following manner: *A motion*[T] *is a geometric transformation*

[T] Isometry or rigid motion. From now on the word "isometry" will be used.

of the plane (or of space) carrying each point A into a new point A′ such that the distance between any two points A and B is equal to the distance between the points A′ and B′ into which they are carried.† However, this definition is rather abstract; now that we realize how basic a role isometries play in geometry, we should like to accept them intuitively and then carefully study all their properties. Such a study is the main content of the first volume of this work. At the end of this volume a complete enumeration of all possible isometries of the plane is given, and this can be taken as a new and simpler definition of them. (For more on this see pages 68–70.)

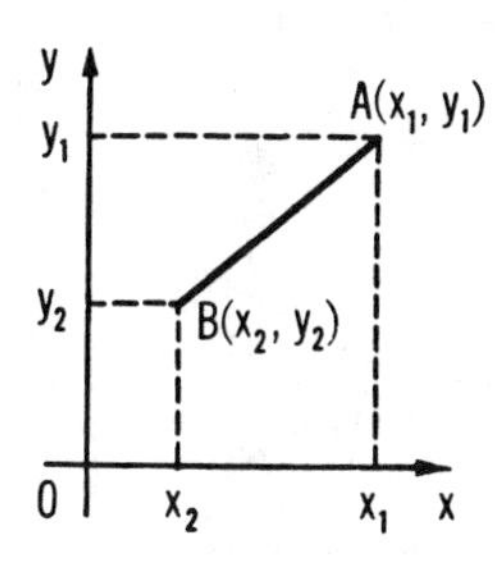

Figure 3

Let us note, moreover, that the study of isometries is essential not only when one wishes to make precise the concepts of geometry, but that it also has a practical importance. The fundamental role of isometries in geometry explains their many applications to the solving of geometric problems, especially construction problems. At the same time the study of isometries provides certain general methods that can be applied to the solution of many geometric problems, and sometimes permits one to combine a series of exercises whose solution by other methods would require

† The distance between two points A and B in the plane is equal to

$$\sqrt{(x_1 - x_2)^2 + (y_1 - y_2)^2}$$

where x_1, y_1 and x_2, y_2 are the coordinates of the points A and B, respectively, in some (it doesn't matter which!) rectangular cartesian coordinate system (Figure 3); thus the concept of distance is reduced to a simple algebraic formula and does not require clarification in what follows.

Analogously, the distance between two points A and B in space is equal to

$$\sqrt{(x_1 - x_2)^2 + (y_1 - y_2)^2 + (z_1 - z_2)^2}$$

where x_1, y_1, z_1 and x_2, y_2, z_2 are the cartesian coordinates of the points A and B in space.

separate consideration. For example, consider the following three well-known problems in construction:

(a) Construct a triangle, given the three points in the plane that are the outer vertices of equilateral triangles constructed outward on the sides of the desired triangle.

(b) Construct a triangle, given the three points in the plane that are the centers of squares constructed outward on the sides of the desired triangle.

(c) Construct a heptagon (polygon of 7 sides), given the seven points that are the midpoints of its sides.

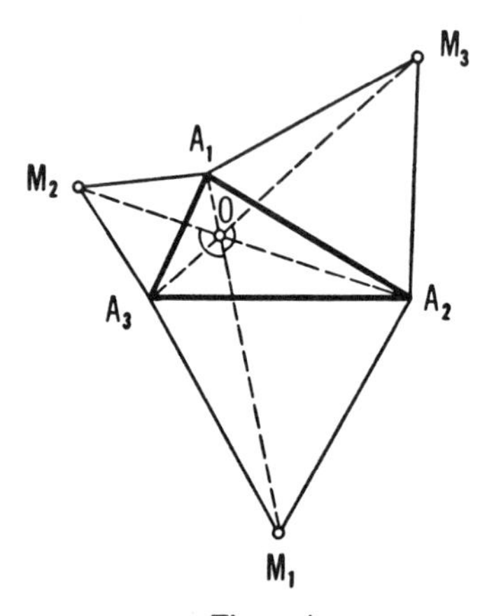

Figure 4a

These problems can be approached with the usual "school book" methods; but then they seem to be three separate problems, independent of one another (and rather complicated problems at that!). Thus the first problem can be solved by proving that the three lines A_1M_1, A_2M_2, and A_3M_3, of Figure 4a, all meet in a point O and form equal angles with one another there (this enables one to find the point O from points M_1, M_2, and M_3, since $\angle M_1OM_2 = \angle M_1OM_3 = \angle M_2OM_3 = 120°$). Then one proves that

$$OA_1 + OA_2 = OM_3, \qquad OA_2 + OA_3 = OM_1 \qquad OA_3 + OA_1 = OM_2$$

[this enables one to find the points A_1, A_2, and A_3 since, for example, $OA_1 = \frac{1}{2}(OM_2 + OM_3 - OM_1)$].

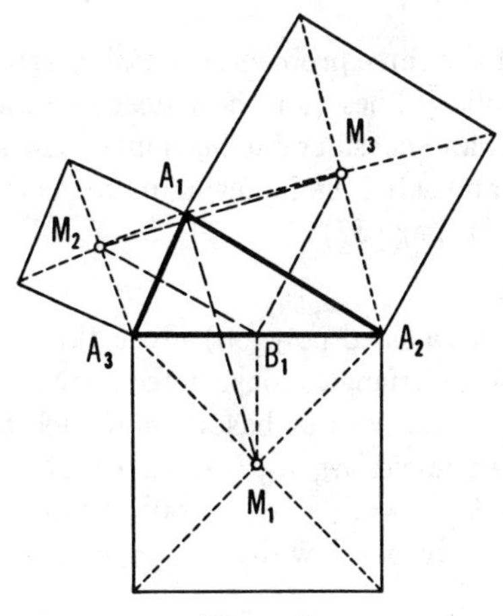

Figure 4b

The second problem can be solved by showing, see Figure 4b, that

$$M_2B_1 \perp M_3B_1 \qquad \text{and} \qquad M_2B_1 = M_3B_1,$$

where B_1 is the midpoint of side A_2A_3 of triangle $A_1A_2A_3$, or (second solution!) that

$$A_1M_1 = M_2M_3 \qquad \text{and} \qquad A_1M_1 \perp M_2M_3.$$

Finally, in solving the third problem one can use the fact that the midpoint M_5' of the diagonal A_1A_5 of the heptagon $A_1A_2A_3A_4A_5A_6A_7$ is the vertex of a parallelogram $M_5M_6M_7M_5'$ (Figure 4c) and therefore can be constructed. Thus we are led to an analogous problem in which the heptagon $A_1A_2A_3A_4A_5A_6A_7$ has been replaced by a pentagon

$$A_1A_2A_3A_4A_5;$$

this new problem can be simplified, again in the same manner.

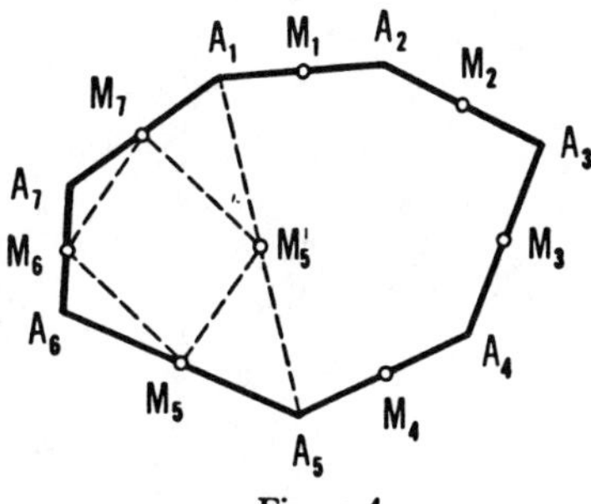

Figure 4c

These solutions of the three problems are rather artificial; they involve drawing certain auxiliary lines (and how does one know which lines to draw?) and they demand considerable ingenuity. The study of isometries enables one to pose and solve the following more general problem in construction (Problem 21, page 37):

Construct an n-gon (n-sided polygon) given the n points that are the outer vertices of isosceles triangles constructed outward on the sides of the desired n-gon (with these sides as bases), and such that these isosceles triangles have vertex angles $\alpha_1, \alpha_2, \cdots, \alpha_n$. [Problem (a) is obtained from this with $n = 3$, $\alpha_1 = \alpha_2 = \alpha_3 = 60°$; Problem (b) with $n = 3$, $\alpha_1 = \alpha_2 = \alpha_3 = 90°$; Problem (c) with $n = 7$, $\alpha_1 = \alpha_2 = \cdots = \alpha_7 = 180°$.]

At the same time this general problem can be solved very simply; with certain general theorems about isometries it can literally be solved in one's head, without drawing any figures. In Chapters 1 and 2 the reader will find a large number of other geometric problems that can be solved with the aid of isometries.

CHAPTER ONE

Displacements

1. Translations

Let us choose a direction NN' in the plane (it may be given, for example, by a line with an arrow); also, let a segment of length a be given. Let A be any point in the plane and let A' be a point such that the segment AA' has the direction NN' and the length a (Figure 5a). In this case we say that the point A' is obtained from the point A by a *translation* in the direction NN' through a distance a, or that the point A is carried into the point A' by this translation. The points of a figure F are carried by the translation into a set of points forming a new figure F'. We say that the new figure F' is obtained from F by a translation (Figure 5b).

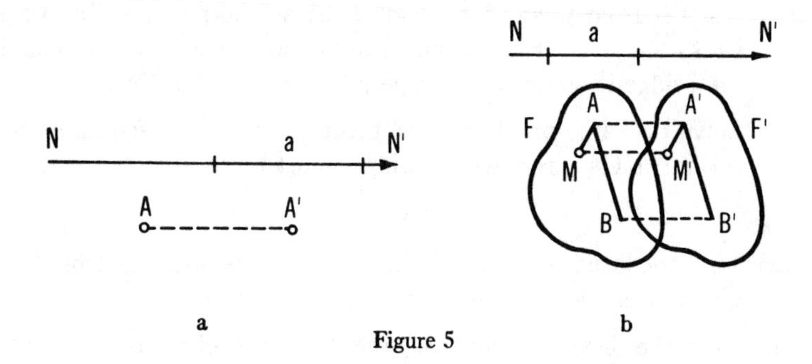

Figure 5

Sometimes we also say that the figure F' is obtained by shifting the figure F "as a whole" in the direction NN' a distance a. Here the expression "as a whole" means that all points of the figure F are moved in the same direction the same distance, that is, that all line segments joining corresponding points in the figures F and F' are parallel, have the same direction, and have the same length. If the figure F' is obtained from F by a translation in the direction NN', then the figure F may be obtained from F' by a translation in the opposite direction to NN' (in the direction $N'N$); this enables us to speak of pairs of figures related by translation.

Translation carries a line l into a parallel line l' (Figure 6a), and a circle S into an equal circle S' (Figure 6b).

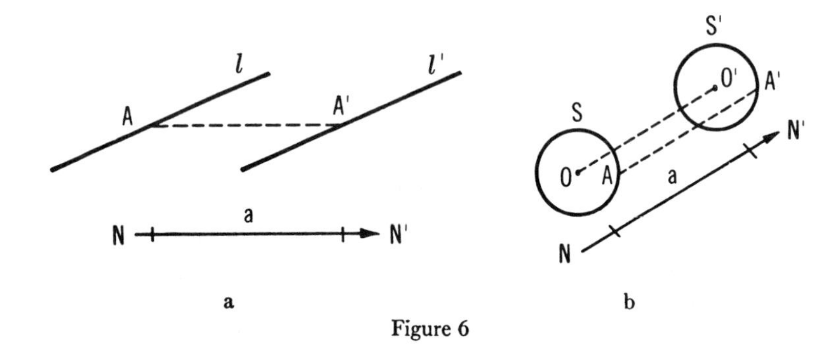

Figure 6

1. Two circles S_1 and S_2 and a line l are given. Locate a line, parallel to l, so that the distance between the points at which this line intersects S_1 and S_2 is equal to a given value a.

2. (a) At what point should a bridge MN be built across a river separating two towns A and B (Figure 7a) in order that the path $AMNB$ from town A to town B be as short as possible (the banks of the river are assumed to be parallel straight lines, and the bridge is assumed to be perpendicular to the river)?

   (b) Solve the same problem if the towns A and B are separated by several rivers across which bridges must be constructed (Figure 7b).

3. (a) Find the locus of points M, the sum of whose distances from two given lines l_1 and l_2 is equal to a given value a.

   (b) Find the locus of points M, the difference of whose distances from two given lines l_1 and l_2 is equal to a given value a.

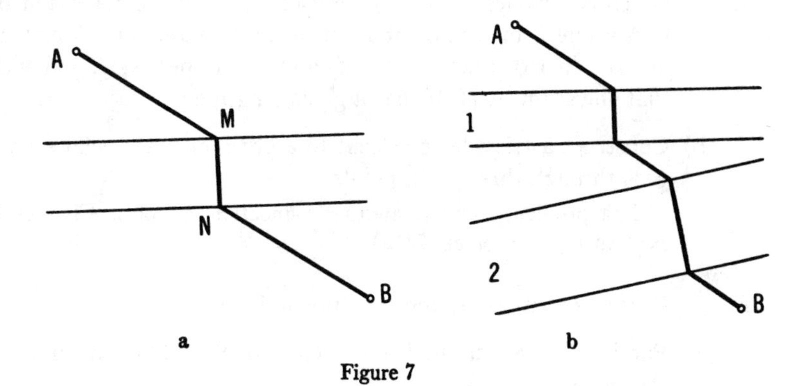

Figure 7

4. Let D, E, and F be the midpoints of sides AB, BC, and CA, respectively, of triangle ABC. Let O_1, O_2, and O_3 denote the centers of the circles circumscribed about triangles ADF, BDE, and CEF, respectively, and let Q_1, Q_2, and Q_3 be the centers of the circles inscribed in these same triangles. Show that the triangles $O_1O_2O_3$ and $Q_1Q_2Q_3$ are congruent.

5. Prove that if the bimedian MN of the quadrilateral $ABCD$ (M is the midpoint of side AD, N is the midpoint of side BC) has length equal to half the sum of the lengths of sides AB and CD, then the quadrilateral is a trapezoid.

6. Given chords AB and CD of a circle; find on the circle a point X such that the chords AX and BX cut off on CD a segment EF having a given length a (Figure 8).

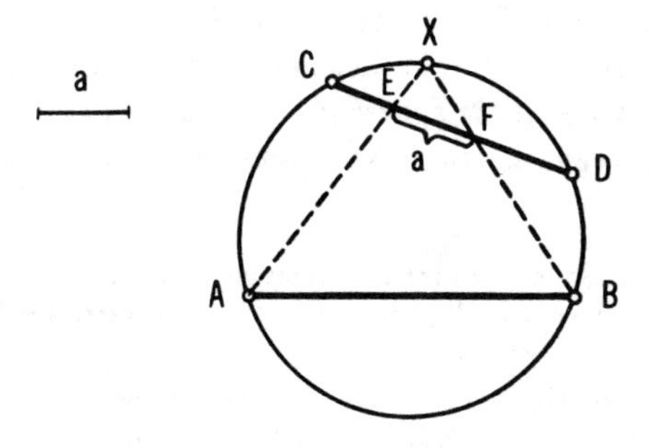

Figure 8

7. (a) Given two circles S_1 and S_2, intersecting in the points A and B; pass a line l through point A, such that it intersects S_1 and S_2 in two other distinct points, M_1 and M_2, respectively, and such that the segment M_1M_2 has a given length a.

   (b) Construct a triangle congruent to a given one, and whose sides pass through three given points.

   This problem occurs in another connection in Vol. 2, Chapter 2, Section 1 [see Problem 73(a)].

8. Given two circles S_1 and S_2; draw a line l:

   (a) Parallel to a given line l_1 and such that S_1 and S_2 cut off equal chords on l.

   (b) Parallel to a given line l_1 and such that S_1 and S_2 cut off chords on l whose sum (or difference) is equal to a given length a.

   (c) Passing through a given point A and such that S_1 and S_2 cut off equal chords on l.

A translation is an example of a transformation of the plane that carries each point A into some other point A'.† Clearly no point is left in place by this transformation; in other words, a translation has no *fixed points* it carries no point into itself.

However there are straight lines that remain in place under a translation; thus, all lines parallel to the direction of the translation are taken into themselves (the lines "slide along themselves"), and therefore these lines (and only these) are *fixed lines* of the translation.

Let us now consider additional properties of translations. Let F and F' be two figures related by a translation; let A and B be any two points of the figure F, and let A' and B' be the corresponding points of the figure F' (see Figure 5b). Since $AA' \parallel BB'$ and $AA' = BB'$,[T] the quadrilateral $AA'B'B$ is a parallelogram; consequently, $AB \parallel A'B'$ and $AB = A'B'$. Thus, *if the figures F and F' are related by a translation, then corresponding segments in these figures are equal, parallel, and have the same direction.*

† This transformation is an *isometry* (motion) in the sense of the definition given in the introduction since, as will presently be shown, it carries each segment AB into a segment $A'B'$ of equal length.

T The statement $AA' = BB'$ means that the lengths of the line segments AA' and BB' are equal. In many books, the distance from a point P to a point Q is denoted by $\overline{PQ}$, but for reasons of typography it will simply be denoted PQ in this book.

Let us show that, conversely, *if to each point of the figure F there corresponds a point of another figure F' such that the segment joining a pair of points in F is equal to, parallel to, and has the same direction as the segment joining the corresponding pair of points in F', then F and F' are related by a translation.* Indeed, choose any pair of corresponding points M and M' of the figures F and F', and let A and A' be any other pair of corresponding points of these figures (see Figure 5b). We are given that $MA \parallel M'A'$ and $MA = M'A'$; consequently the quadrilateral $MM'A'A$ is a parallelogram and, therefore, $AA' \parallel MM'$ and $AA' = MM'$, that is, the point A' is obtained from A by a translation in the direction of the line MM' a distance equal to MM'. But since A and A' were an arbitrary pair of corresponding points, this means that the entire figure F' is obtained from F by a translation in the direction MM' a distance equal to MM'.

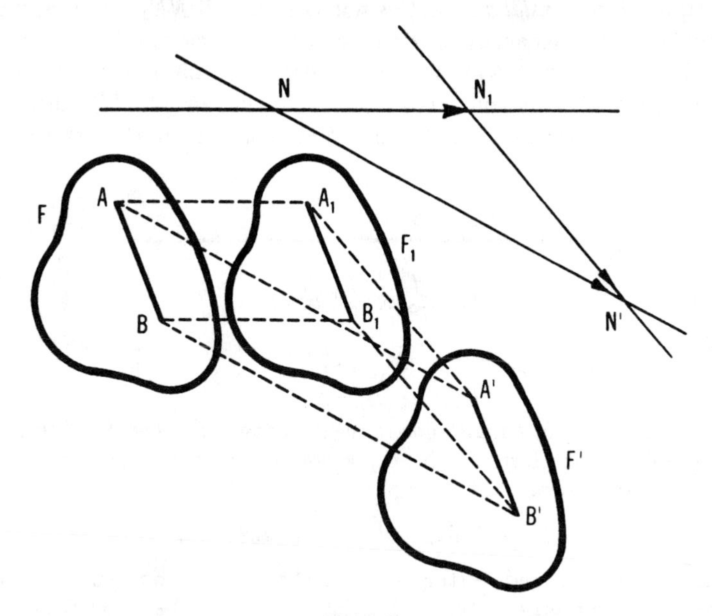

Figure 9

Let us now consider the result of performing *two* translations one after the other. Suppose that the first translation carries the figure F into a figure F_1 and the second carries the figure F_1 into a figure F' (Figure 9). Let us prove that there exists a single translation carrying the figure F into the figure F'. Indeed, if the first translation carries a segment AB of the figure F into the segment A_1B_1 of the figure F_1, then $A_1B_1 \parallel AB$, $A_1B_1 = AB$, and the segments A_1B_1 and AB have the same direction; in exactly the same way the second translation carries A_1B_1 into a segment

$A'B'$ such that $A'B' \parallel A_1B_1$, $A'B' = A_1B_1$ and the segments $A'B'$ and A_1B_1 have the same direction. From this it is clear that corresponding segments AB and $A'B'$ of the figures F and F' are equal, parallel, and have the same direction. But this means that there exists a translation carrying F into F'. Thus, *any sequence of two translations can be replaced by a single translation.*

This last assertion can be formulated differently. In mechanics the replacing of several displacements by a single one, equivalent to all the others, is usually called "addition of the displacements"; in this same sense we shall speak of the *addition of transformations*, where *the sum of two transformations of the plane* is the transformation that is obtained if we first perform one transformation and then perform the second.† Then the result obtained above can be reformulated as follows: *The sum of two translations is a translation.*‡ Let us note also that if NN_1 is the segment that indicates the distance and the direction of the first translation (carrying F into F_1), and if N_1N' is the segment that indicates the distance and direction of the second translation (carrying F_1 into F'), then the segment NN' indicates the distance and direction of the translation carrying F into F'.

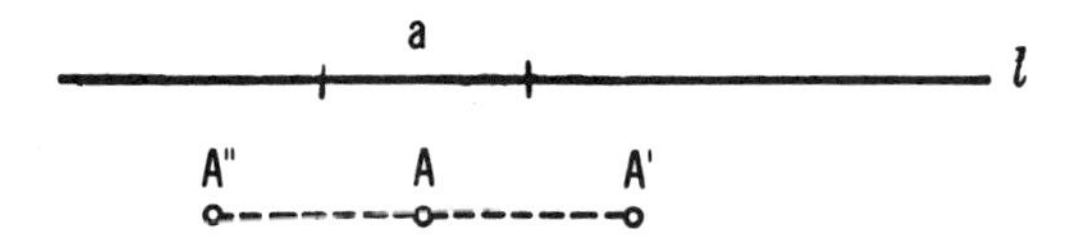

Figure 10

One often speaks of a translation in the direction of a known line l through a given distance a. However, this expression is not exact, since for a given point A the conditions

$$1.\ AA' \parallel l, \qquad 2.\ AA' = a$$

define two points A' and A'' (Figure 10), and not one. In order to make this expression more precise we proceed as follows. One of the directions of the line l is chosen as positive (it may be indicated by an arrow), and the quantity a is considered positive or negative according to whether the direction of the translation coincides with the positive direction of the line l or is opposite to it. Thus the two points A' and A'' in Figure 10 correspond to different

† In mathematical literature the term "product of transformations" is often used in the same sense.

‡ Here is still another formulation of the same proposition: *Two figures F and F' that may each separately be obtained by translation from one and the same third figure F_1 may be obtained from each other by a translation.*

(in sign) distances of translation. Thus the concept of *directed segments* of a line arises naturally; the segments can be positive or negative.

Translation can also be characterized by a single *directed segment* NN' *in the plane*, which indicates at once both the direction and the magnitude of the translation (Figure 11). Thus we are led to the concept of directed line segments (*vectors*) in the plane; these also arise from other considerations in mechanics and physics. Let us note also that the concept of addition of translations leads to the usual definition of addition of vectors (see Figure 9).

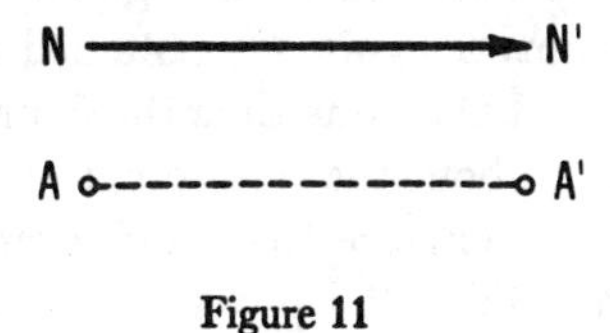

Figure 11

2. Half Turn and Rotation[T]

The point A' is said to be obtained from the point A by means of a half turn about the point O (called the *center of symmetry*) if O is the midpoint of the segment AA' (Figure 12a). Clearly, if the point A' is obtained from A by means of a half turn about O, then also, conversely, A is obtained from A' by means of a half turn about O; this enables one to speak of a pair of points related by a half turn about a given point. If A' is obtained from A by a half turn about O, then one also says that A' *is obtained from A by reflection in the point O*, or that A' *is symmetric to A* with respect to the point O.

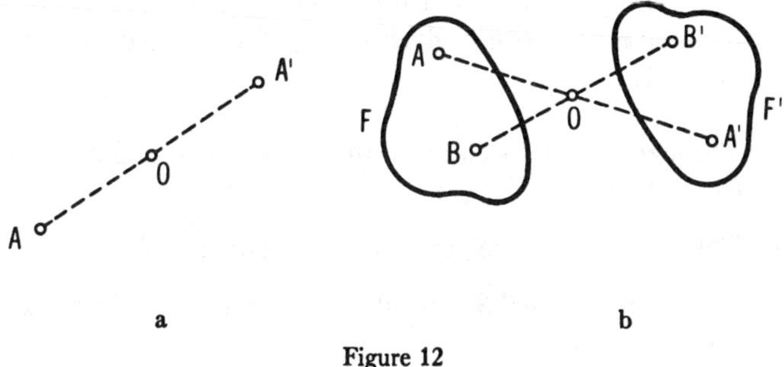

a b

Figure 12

[T] In the original, "half turn" is called "symmetry with respect to a point".

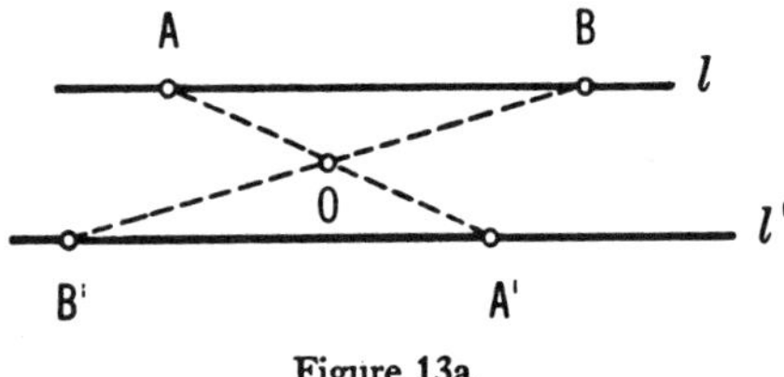

Figure 13a

The set of all points obtained from a given figure F by a half turn about the point O forms a figure F', obtained from F by a half turn about O (Figure 12b); at the same time the figure F is obtained from F' by means of a half turn about the same point O. By a half turn, a line is taken into a parallel line (Figure 13a), and a circle is taken into a congruent circle (Figure 13b). (To prove, for example, that a circle of radius r is taken by a half turn into a congruent circle, it is sufficient to observe that the triangles AOM and $A'OM'$, in Figure 13b, are congruent; consequently, the locus of points A whose distance from M is equal to r is taken into the locus of points A' whose distance from M' is equal to r.)

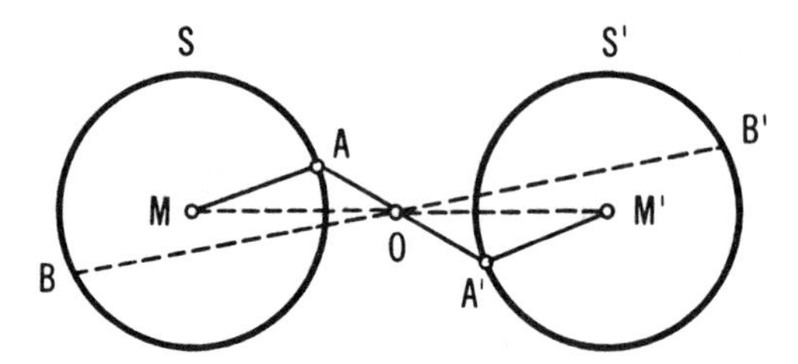

Figure 13b

9. Pass a line through a given point A so that the segment included between its point of intersection with a given line l and its point of intersection with a given circle† S is divided in half by the point A.

10. Through a point A common to two circles S_1 and S_2, pass a line l such that:

(a) The circles S_1 and S_2 cut off equal chords on l.

(b) The circles S_1 and S_2 cut off chords on l whose difference has a given value a.

Problem 10(b) is, clearly, a generalization of Problem 7(a).

† Here we have in mind either one of the points of intersection of the line l with the circle S.

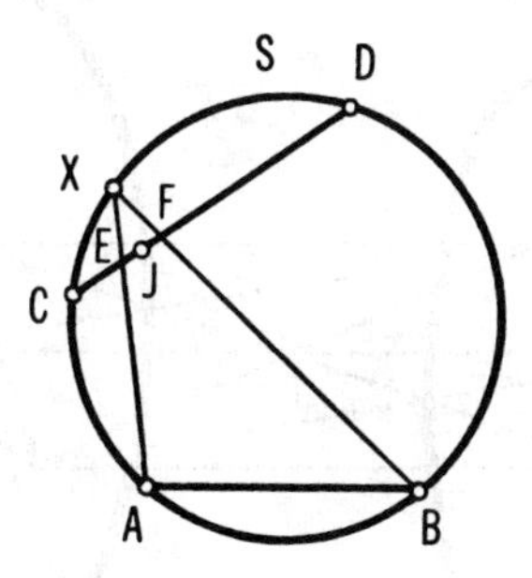

Figure 14

11. Suppose that two chords AB and CD are given in a circle S together with a point J on the chord CD. Find a point X on the circumference, such that the chords AX and BX cut off on the chord CD a segment EF whose midpoint is J (Figure 14).

12. The strip formed by two parallel lines clearly has infinitely many centers of symmetry (Figure 15). Can a figure have more than one, but only a finite number of centers of symmetry (for example, can it have two and only two centers of symmetry)?

Figure 15

If F and F' are two figures related by a half turn about the point O, and if AB and $A'B'$ are corresponding segments of these two figures (Figure 16), then the quadrilateral $ABA'B'$ will be a parallelogram (since its diagonals are divided in half by their point of intersection O). From this it is clear that *corresponding segments of two figures related by a half turn about a point are equal, parallel, and oppositely directed.* Let us show that, conversely, *if to each point of a figure F one can associate a point of a figure F' such that the segments joining corresponding points of these figures are equal, parallel, and oppositely directed, then F and F' are related by a half turn about some point.* Indeed, choose a pair of corresponding points M and M' of the figures F and F' and let O be the midpoint of the segment MM'. Let A, A' be any other pair of corresponding points

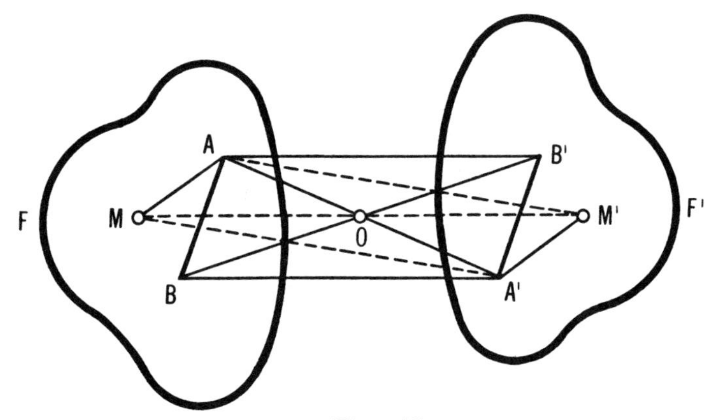

Figure 16

of these figures (see Figure 16). We are given that $AM \parallel M'A'$ and $AM = M'A'$; consequently the quadrilateral $AMA'M'$ is a parallelogram and, therefore, the midpoint of the diagonal AA' coincides with the midpoint O of the diagonal MM'; that is, the point A' is obtained from A by a half turn about the point O. And since the points A and A' were an arbitrary pair of corresponding points, it follows that the figure F' is obtained from F by a half turn about O.

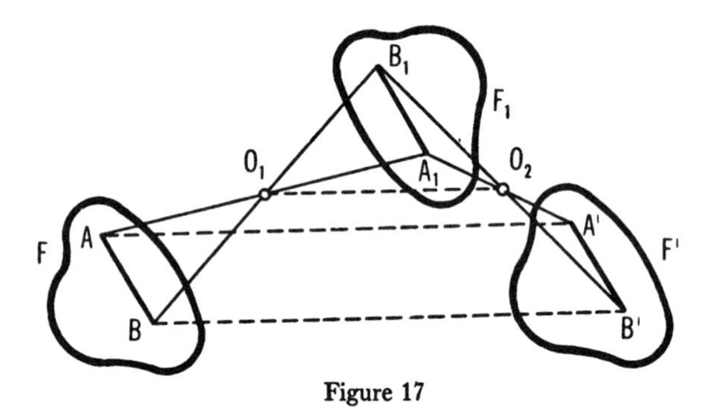

Figure 17

Let us now consider three figures F, F_1, and F' such that the figure F_1 is obtained from F by a half turn about the point O_1, and the figure F' is obtained from F_1 by a half turn about the point O_2 (Figure 17). Let A_1B_1 be an arbitrary segment of the figure F_1, and let AB and $A'B'$ be the corresponding segments of the figures F and F'. Then the segments A_1B_1 and AB are equal, parallel, and oppositely directed; the segments A_1B_1 and $A'B'$ are also equal, parallel, and oppositely directed. Consequently the segments AB and $A'B'$ are equal, parallel, and have the same direction. But once corresponding segments of the figures F and F' are equal, parallel, and have the same direction, then *F' may be obtained from F by means of a translation.* Thus *the sum of two half turns is a translation* (compare above, page 20). This can also be seen directly from Figure 17. Since O_1O_2 is a line joining the midpoints of the sides AA_1 and $A'A_1$ of the triangle AA_1A', it follows that $AA' \parallel O_1O_2$ and $AA' = 2O_1O_2$; that is, *each point A' of the figure F' is obtained from the corresponding point A of the figure F by a translation in the direction O_1O_2 through a distance equal to twice the segment O_1O_2.*

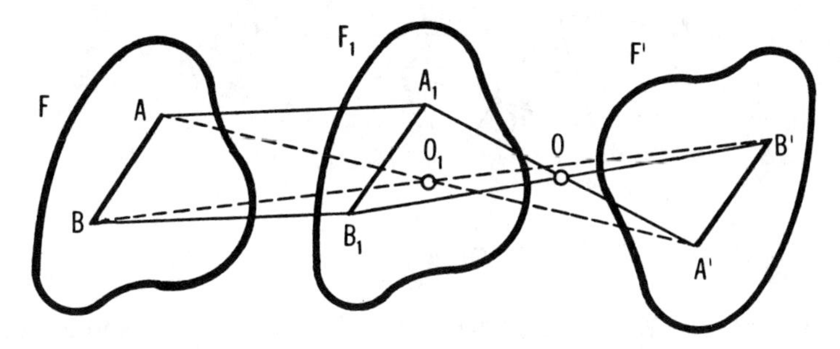

Figure 18

In exactly the same way it can be shown that *the sum of a translation and a half turn about a point O* (Figure 18), *or of a half turn and a translation, is a half turn about some new point O_1.*

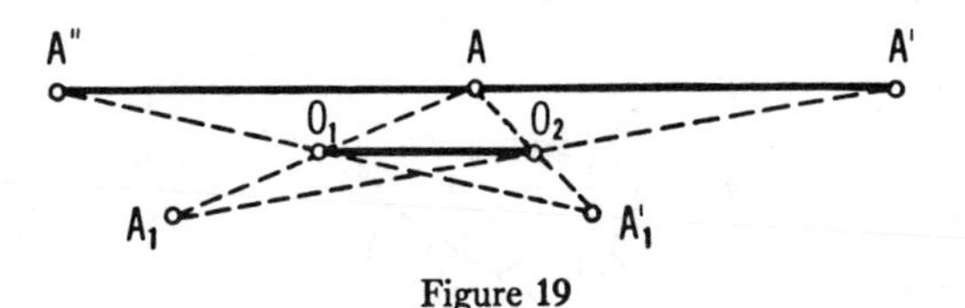

Figure 19

Let us make one further important observation. The sequence of half turns about the point O_1 and O_2 (in Figure 19: $A \to A_1 \to A'$) is equivalent to a translation of distance $2O_1O_2$ in the direction from O_1 to O_2, while the sequence of these same half turns, carried out in the reverse order (in Figure 19: $A \to A_1' \to A''$), is equivalent to a translation of the same distance in the direction from O_2 to O_1. Thus, the sum of two half turns depends in an essential way on the order in which these half turns are performed. This circumstance is, in general, characteristic of the addition of transformations: *The sum of two transformations depends, in general, on the order of the terms.*

In speaking of the addition of half turns, we considered the half turn as a transformation of the plane, carrying each point A into a new point A'.† It is not difficult to see that *the only point left fixed by a half turn is the center O about which the half turn is taken,* and that *the fixed lines are the lines that pass through the center O.*

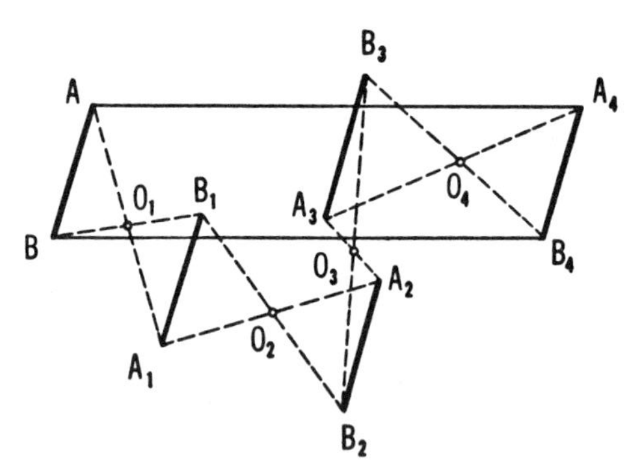

Figure 20a

13. (a) Let $O_1, O_2, \cdots, O_n$ (n even) be points in the plane and let AB be an arbitrary segment; let the segment A_1B_1 be obtained from AB by a half turn about O_1, let A_2B_2 be obtained from A_1B_1 by a half turn about O_2, let A_3B_3 be obtained from A_2B_2 by a half turn about $O_3, \cdots$, finally, let A_nB_n be obtained from $A_{n-1}B_{n-1}$ by a half turn about O_n (see Figure 20a, where $n = 4$). Show that $AA_n = BB_n$.

Does the assertion of this exercise remain true if n is odd?

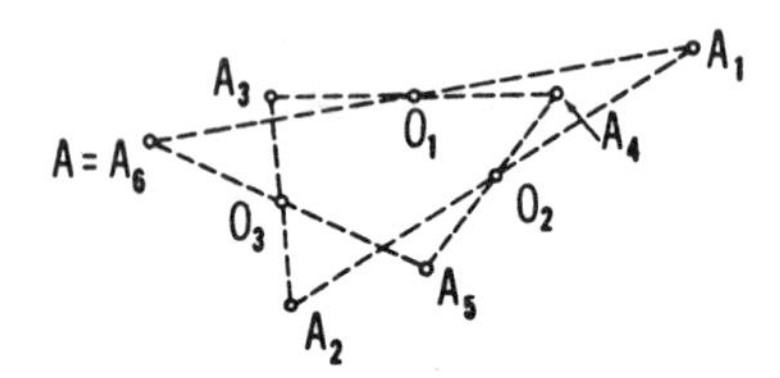

Figure 20b

(b) Let an odd number of points $O_1, O_2, \cdots, O_n$ be given in the plane (see Figure 20b, where $n = 3$). Let an arbitrary point A be moved successively by half turns about $O_1, O_2, \cdots, O_n$ and then once again moved successively by half turns about the

† See the footnote † on page 18.

same points O_1, $O_2 \cdots$, O_n. Show that the point A_{2n}, obtained as the result of these $2n$ half turns, coincides with the point A.

Does the assertion of the problem remain true if n is even?

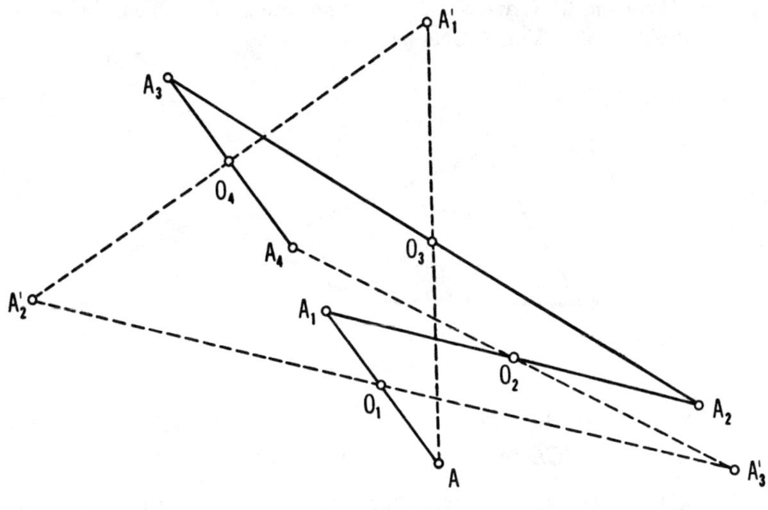

Figure 21

14. (a) Let O_1, O_2, O_3, O_4 be any four points in the plane. Let an arbitrary fifth point A be moved successively by half turns about the points O_1, O_2, O_3, O_4. Now starting again with the original point A, let it be moved successively by half turns about the same four points, but in the following order: O_3, O_4, O_1, O_2. Show that in both cases the position of the final point A_4 is the same (see Figure 21).

(b) Let O_1, O_2. O_3, O_4, O_5 be any five points in the plane. Let an arbitrary point A be moved by successive half turns about these five points. Now starting again with the original point A, let it be moved by successive half turns about these same five points taken in reverse order: O_5, O_4, O_3, O_2, O_1. Show that in both cases the position of the final point A_5 is the same.

(c) Let n points O_1, O_2, $\cdots$, O_n be given in the plane. An arbitrary point is moved by successive half turns about the points O_1, O_2, $\cdots$, O_n; then the same original point is moved successively by half turns about these same points taken in reverse order: O_n, O_{n-1}, $\cdots$, O_1. For which values of n will the final positions be the same in both cases?

15. Let n be an odd number (for example, $n = 9$), and let n points be given in the plane. Find the vertices of an n-gon that has the given points as midpoints of its sides.

Consider the case when n is even.

Problem 21 (page 37) is a generalization of Problem 15, as is Problem 66 of Vol. 2, Chapter 1, Section 2.

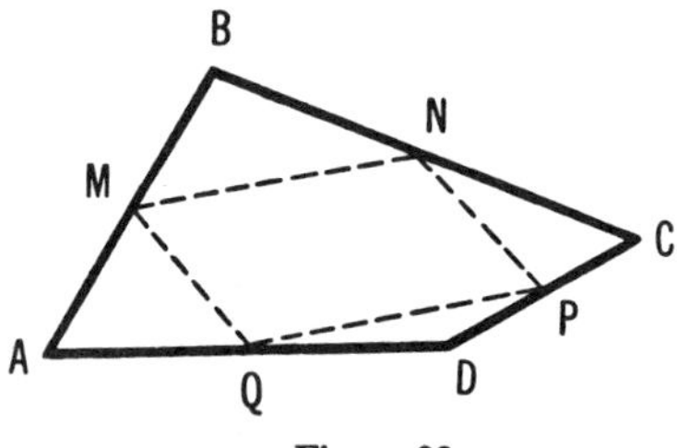

Figure 22a

16. (a) Prove that the midpoints of the sides of an arbitrary quadrilateral $ABCD$ form a parallelogram (Figure 22a).

(b) Let M_1, M_2, M_3, M_4, M_5, M_6 be the midpoints of the sides of an arbitrary hexagon. Prove that there exists a triangle T_1 whose sides are equal and parallel to the segments M_1M_2, M_3M_4, M_5M_6, and a triangle T_2 whose sides are equal and parallel to M_2M_3, M_4M_5, M_6M_1 (Figure 22b).

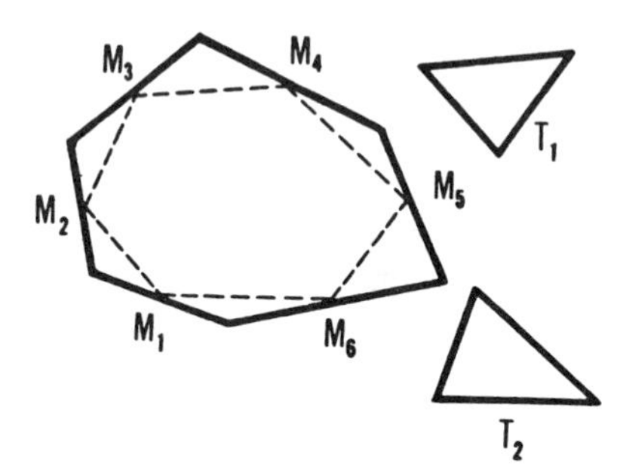

Figure 22b

Choose in the plane a point O; let an angle α be given and let us agree on a direction of rotation (we shall assume, for example, that it is opposite to the direction in which the hands of a clock move). Let A be an arbitrary point of the plane and let A' be the point such that $OA' = OA$ and

$\measuredangle AOA' = \alpha$ (so that OA must be turned through an angle α in the direction we have chosen in order to coincide with OA'). In this case we say that *the point A' is obtained from the point A by means of a rotation with center O and angle of turning α*, or that the point A is carried into A' by this rotation (Figure 23a). The set of all points obtained from points of a figure F by a rotation about a point O through an angle α forms a new figure F' (Figure 23b). Sometimes one says that the figure F' is obtained by rotating the figure F "as a whole" about the point O through an angle α; here the words "as a whole" mean that all points of the figure F are moved along circles with one and the same center O and that they all describe the same arcs (in angular measure) of these circles. If the figure F' is obtained by a rotation from the figure F, then, conversely, the figure F may be obtained from the figure F' by a rotation with the same center and with angle of rotation $360°—\alpha$ (or by a rotation through the same angle α, but in the opposite direction); this permits one to speak of pairs of figures obtained from each other by rotation.

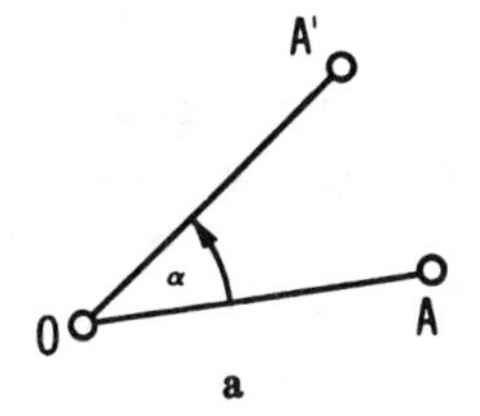

a

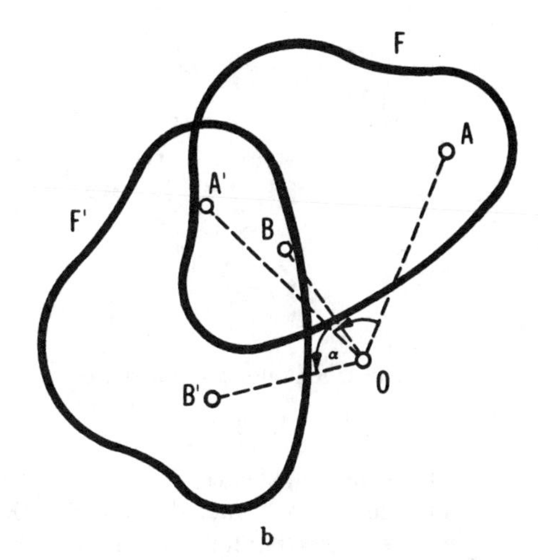

b

Figure 23

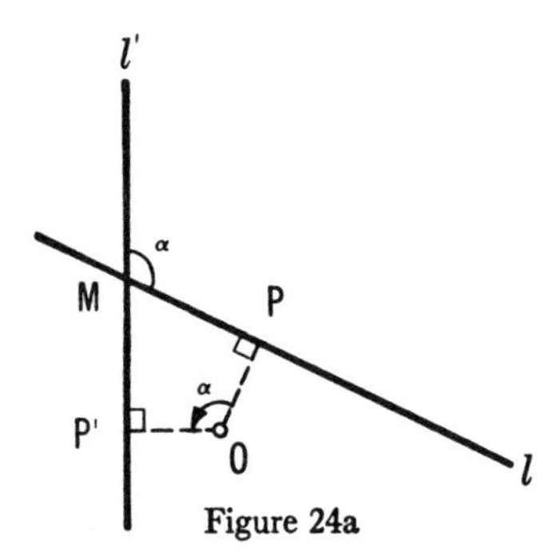

Figure 24a

A line l is taken by a rotation about a point O into a new line l'; in order to find l' it is sufficient to rotate the foot P of the perpendicular from O to l, and then to pass a line through the new point P' perpendicular to OP' (Figure 24a). Clearly the angle α between the lines l and l' is equal to the angle of rotation; to prove this it is sufficient to observe that the angles POP' and lMl', in Figure 24a, are equal because they are angles with mutually perpendicular sides.

A circle S is taken into a new circle S' by a rotation about a point O; to construct S' one must rotate the center M of the circle S about O and then construct a circle with the new point M' as center and with the same radius as the original circle S (Figure 24b).

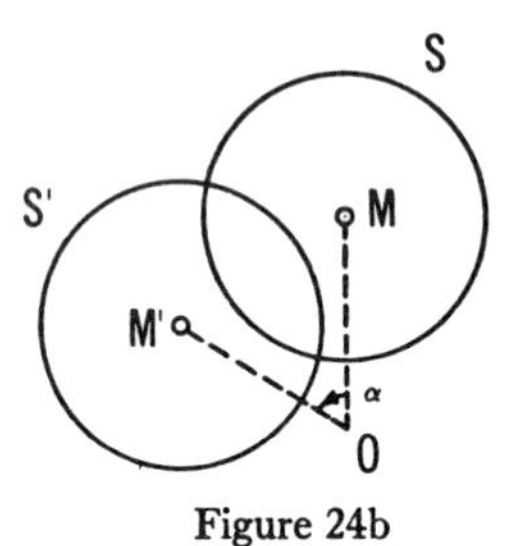

Figure 24b

Clearly when one is given the point A, the conditions

$$1.\quad OA' = OA, \qquad\qquad 2.\quad \sphericalangle AOA' = \alpha,$$

without any supplementary agreement on the direction of rotation, determine two points A' and A'' (Figure 25). To select one of them we may, for example, proceed as follows. Let us agree to consider one direction of rotation as positive (it can be indicated, for example, by an arrow on a circle), and the opposite direction as negative. Further, we shall consider the angle of rotation $\alpha = \measuredangle AOA'$ as positive or negative, depending on the direction of the rotation carrying A into A'; in this case the two points A' and A'' will correspond to different angles of rotation (differing in sign). Thus we are naturally led to the concept of *directed angles* that can be positive as well as negative; this concept is useful in many other questions of elementary mathematics. (The concept of directed circles, that is, circles on which some direction has been chosen, also arises in other connections.)

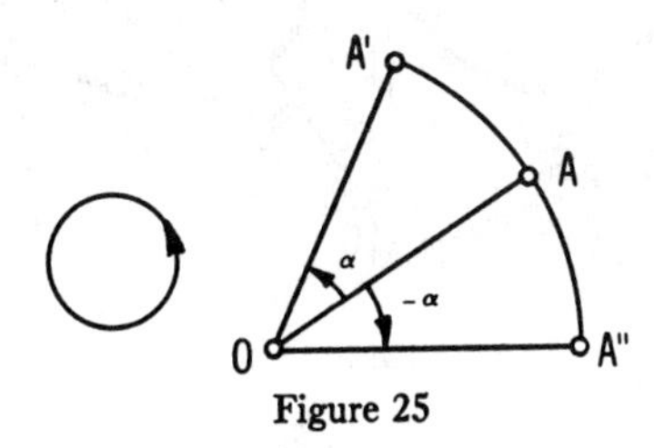

Figure 25

17. Let two lines l_1 and l_2, a point A, and an angle α be given. Find a circle with center A such that l_1 and l_2 cut off an arc whose angular measure is equal to α.

18. Find an equilateral triangle whose vertices lie on three given parallel lines or on three given concentric circles.

19. Let a circle S, points A and B, and an angle α be given. Find points C and D on S such that $CA \parallel DB$ and arc $CD = \alpha$.

20. Let two circles S_1 and S_2, a point A, and an angle α be given. Pass lines l_1 and l_2 through A forming an angle α, such that the circles S_1 and S_2 cut off equal chords on these lines.

Let the rotation with center at the point O and angle of rotation α carry the figure F into the figure F', and let AB and $A'B'$ be corresponding segments of these figures (Figure 26). Then the triangles OAB and $OA'B'$ are congruent ($OA = OA'$, $OB = OB'$ and $\measuredangle AOB = \measuredangle A'OB'$, since $\measuredangle AOA' = \measuredangle BOB' = \alpha$); consequently, $AB = A'B'$. The angle between the segments AB and $A'B'$ is equal to α (because the lines AB and $A'B'$ are related by a rotation through an angle α; see Figure 24a); at the same time we must turn AB through an angle α in the direction

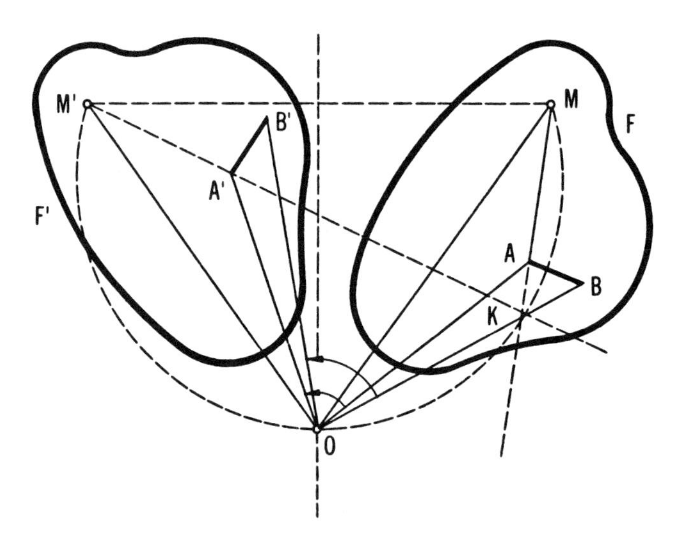

Figure 26

of the rotation, in order to obtain the directed segment $A'B'$.† Thus we see that *if the figures F and F′ are related by a rotation through an angle α, then corresponding segments of these figures are equal and make an angle α with each other.*

Let us show that, conversely, *if to each point of the figure F there corresponds a point of another figure F′, and these figures are such that corresponding segments are equal and make an angle α with each other* (so that the segments of the figure F become parallel to the corresponding segments of the figure F' when they are turned through an angle α in the chosen direction), *then F and F′ are related by a rotation through an angle α about some center.* Indeed, let M and M' be two corresponding points of the figures F and F'. Construct on the segment MM' a circle arc[T] subtending an angle α, and let O be the point of intersection of this arc with the perpendicular to the segment MM' at its midpoint. Since $OM = OM'$

† The angle between two segments AB and $A'B'$ that do not pass through a common point is by definition the angle between the lines through AB and $A'B'$. This is the angle through which we must turn AB in order to make it parallel to the segment $A'B'$.

From this last remark it follows that *if one has three segments AB, A_1B_1, and $A'B'$, then the angle between the first and third is equal to the sum of the angles between the first and second and between the second and third.* (To be completely accurate one should speak of directed angles; see the small print on page 30.) We shall soon use this fact.

[T] For the details of this construction, see, for example, *Hungarian Problem Book 1* in this series, Problem 1895/2, Note.

and $\sphericalangle MOM' = \alpha$, it follows that the rotation with center O and angle α carries the point M into M'.† Further, let A and A' be any other corresponding points of the figures F and F'. Consider the triangles OMA and $OM'A'$. One has $OM = OM'$ (by construction of the point O), $MA = M'A'$ (this was given); in addition, $\sphericalangle OMA = \sphericalangle OM'A'$, because the angle between OM and OM' is equal to the angle between MA and $M'A'$, that is, the points M, M', O, and K (K is the point of intersection of AM and $A'M'$) lie on a circle and the inscribed angles OMA and $OM'A'$ cut off the same arc. Therefore the triangles OMA and $OM'A'$ are congruent. From this it follows that $OA = OA'$; moreover, $\sphericalangle AOA' = \sphericalangle MOM' = \alpha$ (because $\sphericalangle A'OM' = \sphericalangle AOM$). Consequently the rotation with center O and angle α carries each point A of the figure F into the corresponding point A' of the figure F', which was to be proved.

Now we are in a position to answer the question: What is represented by the sum of two rotations? First of all, it is clear from the very definition of rotation that the sum of two rotations (in the same direction or sense) with common center O and with angles of rotation respectively equal to α and β is a rotation about the same center O with angle of rotation $\alpha + \beta$ (Figure 27a).

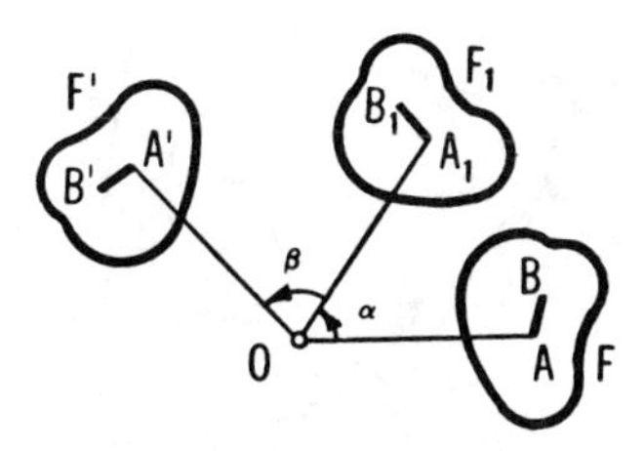

Figure 27a

Now let us consider the general case. Let the figure F_1 be obtained from F by means of a rotation with center O_1 and angle α, and let the figure F' be obtained from F_1 by a rotation in the same sense with center

† The conditions $OM = OM'$ and $\sphericalangle MOM' = \alpha$ define two points O (the perpendicular can be constructed on either side of MM'). We must choose one of these two points such that the direction of the rotation with center O carrying M into M' coincides with the direction of the rotation through an angle α that carries segments of the figure F into positions parallel to the corresponding segments of the figure F'.

O_2 and angle β (Figure 27b). If the first rotation carries the segment AB of the figure F into the segment A_1B_1 of the figure F_1, and if the second rotation carries the segment A_1B_1 into the segment $A'B'$ of the figure F', then the segments AB and A_1B_1 are equal and form an angle α; the segments A_1B_1 and $A'B'$ are equal and form an angle β. Thus corresponding segments AB and $A'B'$ of the figures F and F' are equal and form an angle $\alpha + \beta$; if $\alpha + \beta = 360°$ this means that corresponding segments of the figures F and F' are parallel.† From this it follows, by what has been proved before, that the figures F and F' are related by a rotation through the angle $\alpha + \beta$, if $\alpha + \beta \neq 360°$, and by a translation if $\alpha + \beta = 360°$. Thus *the sum of two rotations in the same sense, with centers O_1 and O_2 and angles α and β is a rotation through the angle $\alpha + \beta$, if $\alpha + \beta \neq 360°$, and is a translation, if $\alpha + \beta = 360°$.* Since a rotation through an angle α is equivalent to a rotation of $360° - \alpha$ in the opposite sense, the last part of the theorem that has been proved may also be reformulated as follows: *The sum of two rotations is a translation if these rotations have the same angles of rotation but opposite directions of rotation.*

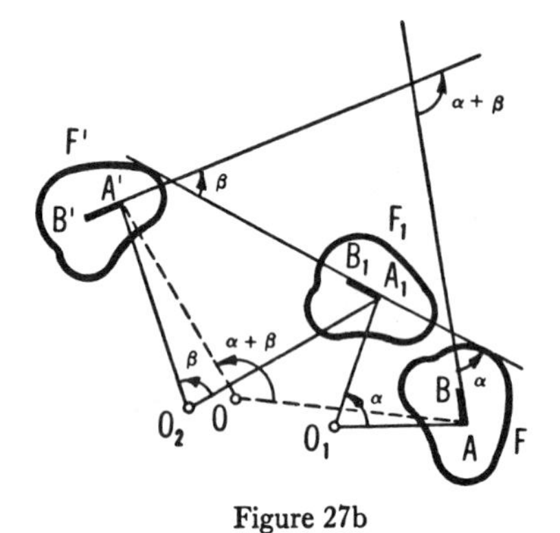

Figure 27b

Let us now show how, from the centers O_1 and O_2 and from the angles α and β of two rotations, one can find the rotation or translation that represents their sum. Suppose first that $\alpha + \beta \neq 360°$. In this case the sum of the rotations is a rotation through the angle $\alpha + \beta$; let us find its

† Strictly speaking we should say that corresponding segments of the figures F and F' are parallel if $\alpha + \beta$ is a multiple of 360°. However we can always assume that α and β are less than 360°; in this case $\alpha + \beta$ is a multiple of 360° only if $\alpha + \beta = 360°$.

center. The sum of the two rotations carries the center O_1 of the first into a point O_1' such that

$$O_1'O_2 = O_1O_2 \quad \text{and} \quad \sphericalangle O_1O_2O_1' = \beta.$$

(See Figure 28a; the first rotation leaves O_1 in place, and the second carries O_1 into O_1'.) The sum of the two rotations carries a point O_2'' into O_2 such that

$$O_2''O_1 = O_2O_1 \quad \text{and} \quad \sphericalangle O_2''O_1O_2 = \alpha$$

(the first rotation carries O_2'' into O_2 and the second leaves O_2 in place). From this it follows that the center O that we are seeking is equidistant from O_2 and O_2'' and from O_1' and O_1; consequently it can be found as the point of intersection of the perpendicular bisectors l_1 and l_2 of the segments O_2O_2'' and $O_1'O_1$ respectively. But from Figure 28a, it is clear that l_1 passes through O_1 and $\sphericalangle l_1O_1O_2 = \frac{1}{2}\alpha$, and that l_2 passes through O_2 and $\sphericalangle O_1O_2l_2 = \frac{1}{2}\beta$. The lines l_1 and l_2 are completely determined by these conditions; we find the desired center of rotation O as their point of intersection.

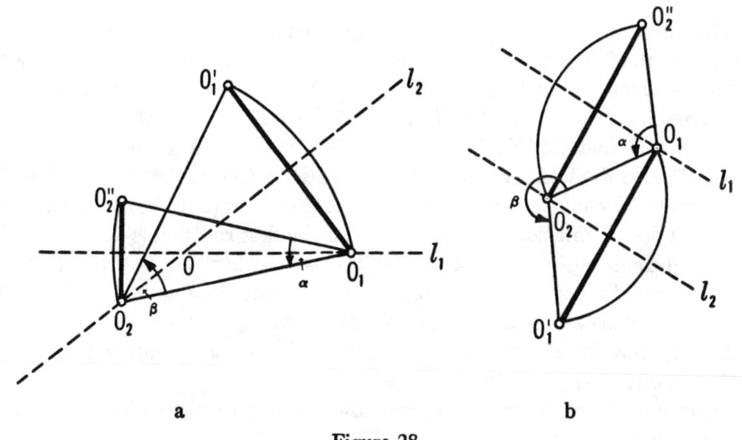

Figure 28

If $\alpha + \beta = 360°$, then the translation that is equal to the sum of the rotations may be determined by the fact that it carries the point O_1 into O_1' (or O_2'' into O_2); here the points O_1' and O_2'' are defined just as before (see Figure 28b; from the picture it is clear that the lines l_1 and l_2 that figured in the previous construction are now parallel—they are perpendicular to the direction of the translation, and the distance between them is half the distance of the translation).

Analogously to the proof of the theorem on the sum of two rotations, it can be shown that *the sum of a translation and a rotation* (*and the sum of a rotation and a translation*) *is a rotation through the same angle as the first rotation, but with a different center.* We shall leave it to the reader to find for himself the construction of the center O_1 of this rotation, given the center O and angle α of the original rotation and the distance and direction of the translation (see also the text printed in small type that follows, and on page 51).

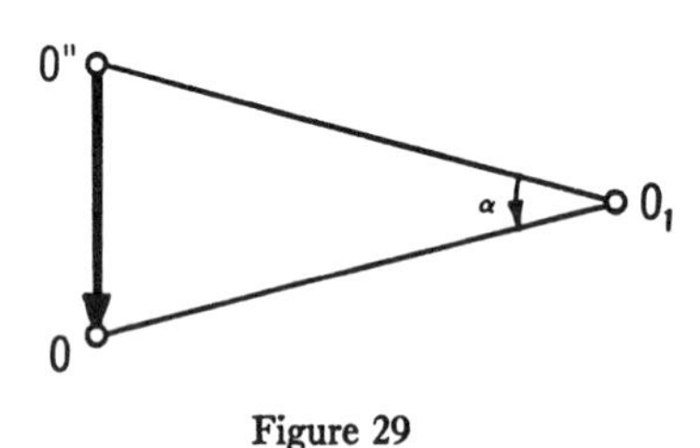

Figure 29

The theorem on the addition of a translation and a rotation can also be proved in the following manner. We know that the sum of two rotations with the same angle α but with opposite directions of rotation is a translation· it carries into the center O_2 of the second rotation a point O_2'' such that $O_1O_2'' = O_1O_2$ and $\measuredangle O_2''O_1O_2 = \alpha$ (see Figure 28b). Let us represent the given translation in the form of a sum of two rotations, the second of which has the same center O and the same angle α as the given rotation, but has the opposite direction of rotation. (The center O_1 of the first rotation is determined by the conditions $O_1O'' = O_1O$ and $\measuredangle O''O_1O = \alpha$, where O'' is a point that is carried into the point O by the given translation; see Figure 29.) Thus the sum of the translation and the rotation has been replaced by the sum of three rotations. But the last two of these rotations annul one another and we are left with one unique rotation with center O_1.

In an analogous way one can prove the theorem on the addition of a rotation and a translation.

One is struck by the great similarity between the properties of rotations and the properties of translations that can be seen by comparing the proofs of the theorems on the addition of translations and on the addition of rotations.† Translation and rotation together are called *displacements* (or *proper motions* or *direct isometries*); the reasons for this name will be explained in Chapter 2, Section 2 (see page 66).

† From a more advanced point of view translation can even be considered as a special case of rotation.

Half turn is a special case of rotation, corresponding to the angle $\alpha = 180°$. We obtain another special case by putting $\alpha = 360°$. A rotation with angle $\alpha = 360°$ returns each point of the plane to its original position; this transformation, in which no point of the plane changes its position, is called the *identity* (or the *identity transformation*). (It may seem that the very word, "transformation", is out of place here, since in the identity transformation all figures remain unchanged; however, this name will be convenient for us.)

Just as in the case of a half turn, a rotation can be regarded as a transformation of the whole plane, carrying each point A into a new point A'.† *The only fixed point of this transformation is the center of rotation O* (the only exception is the case when the angle of rotation α is a multiple of 360°, that is, when the rotation is the identity); *a rotation has no fixed lines at all* (except when α is a multiple of 180°, that is, when the rotation is either the identity or a half turn).

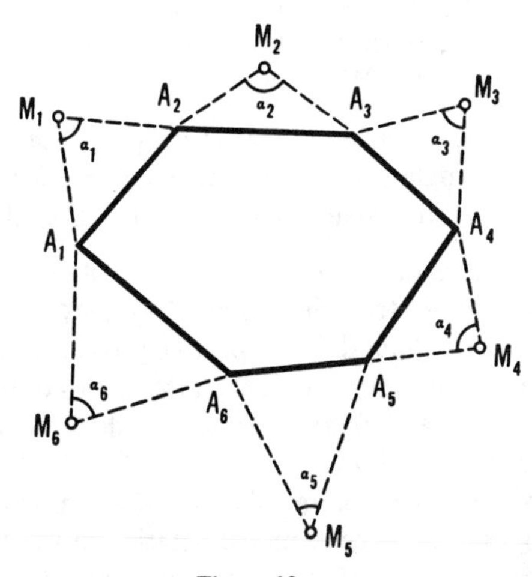

Figure 30

21. Construct an n-gon, given the n points that are the vertices of isosceles triangles constructed on the sides of the n-gon, with the angles $\alpha_1, \alpha_2, \cdots, \alpha_n$ at the outer vertices (see Figure 30, where $n = 6$).

Problem 15 is a special case of Problem 21 (there n is odd and $\alpha_1 = \alpha_2 = \cdots = \alpha_n = 180°$). Problem 66 of Vol. 2, Chapter 2 is a generalization of Problem 21.

† See the footnote † on page 18.

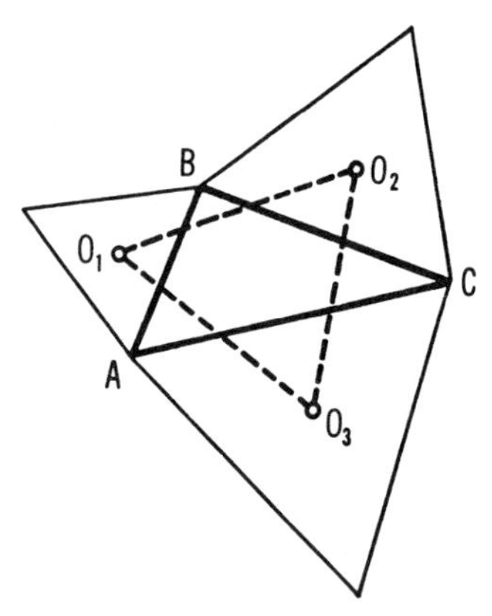

Figure 31

22. (a) Construct equilateral triangles on the sides of an arbitrary triangle ABC, exterior to it. Prove that the centers O_1, O_2, O_3 of these triangles themselves form the vertices of an equilateral triangle (Figure 31).

   Does the assertion of this exercise remain correct if the equilateral triangles are not constructed exterior to triangle ABC, but on the same side of its sides as the triangle itself?

   (b) On the sides of an arbitrary triangle ABC, exterior to it, construct isosceles triangles BCA_1, ACB_1, ABC_1 with angles at the vertices A_1, B_1, and C_1, respectively equal to α, β, and γ. Prove that if $\alpha + \beta + \gamma = 360°$, then the angles of the triangle $A_1B_1C_1$ are equal to $\frac{1}{2}\alpha$, $\frac{1}{2}\beta$, $\frac{1}{2}\gamma$, that is, they do not depend on the shape of the triangle ABC.

   Does the assertion of this exercise remain valid if the isosceles triangles are not constructed exterior to the triangle ABC, but on the same side of its sides as the triangle itself?

   It is not difficult to see that Problem 22(a) is a special case of Problem 22(b) (with $\alpha = \beta = \gamma = 120°$).

23. On the sides of an arbitrary triangle ABC construct equilateral triangles BCA_1, ACB_1, and ABC_1, so that the vertices A_1 and A are on opposite sides of BC, B_1 and B are on opposite sides of AC, but C_1 and C are on the same side of AB. Let M be the center of triangle ABC_1. Prove that A_1B_1M is an isosceles triangle with an angle of 120° at the vertex M (Figure 32).

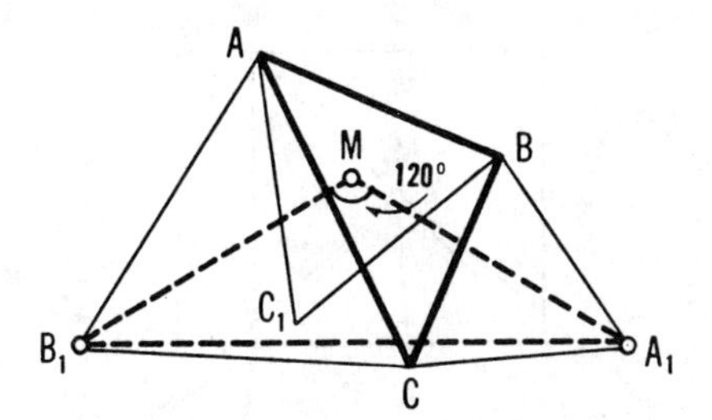

Figure 32

24. (a) On the sides of an arbitrary (convex) quadrilateral $ABCD$ equilateral triangles ABM_1, BCM_2, CDM_3, and DAM_4 are constructed, so that the first and third of them are exterior to the quadrilateral, while the second and fourth are on the same side of sides BC and DA as is the quadrilateral itself. Prove that the quadrilateral $M_1M_2M_3M_4$ is a parallelogram (see Figure 33a; in special cases this parallelogram may degenerate into an interval).

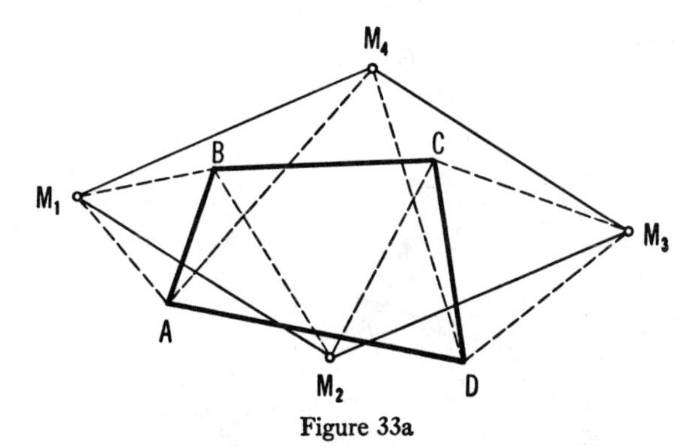

Figure 33a

(b) On the sides of an arbitrary (convex) quadrilateral $ABCD$ squares are constructed, all lying exterior to the quadrilateral; the centers of these squares are M_1, M_2, M_3, and M_4. Show that $M_1M_3 = M_2M_4$ and $M_1M_3 \perp M_2M_4$ (Figure 33b).

(c) On the sides of an arbitrary parallelogram $ABCD$ squares are constructed, lying exterior to it. Prove that their centers M_1, M_2, M_3, M_4 are themselves the vertices of a square (Figure 33c).

Is the assertion of this problem still correct if the squares all lie on the same side of the sides of the parallelogram as does the parallelogram itself?

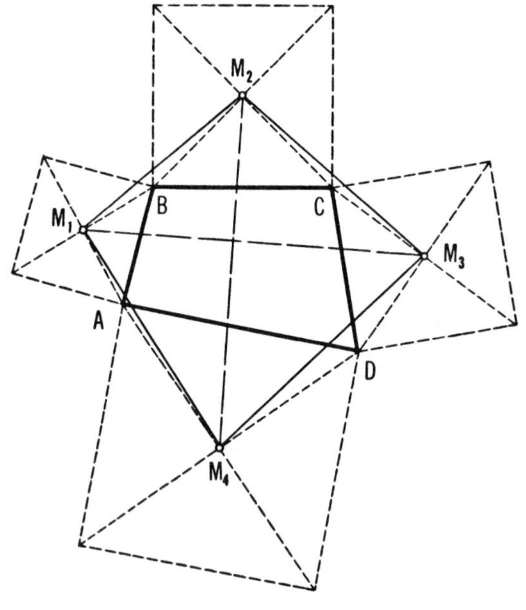

Figure 33b

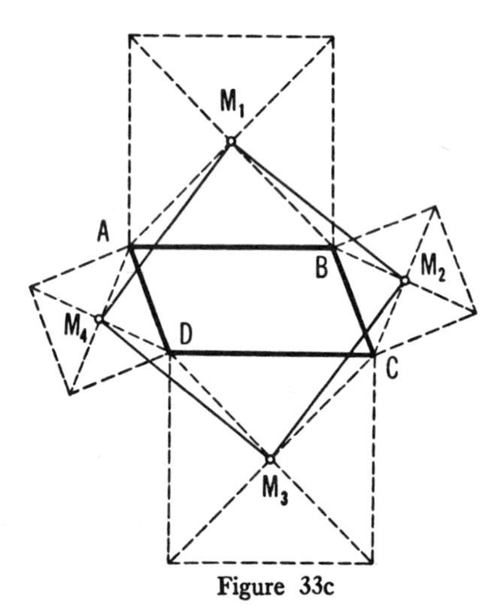

Figure 33c

CHAPTER TWO

Symmetry

1. Reflection and Glide Reflection

A point A' is said to be the image of a point A by reflection in a line l (called the axis of symmetry) if the segment AA' is perpendicular to l and is divided in half by l (Figure 34a). If the point A' is the image of A in l, then, conversely, A is the image of A' in l; this enables one to speak of pairs of points that are images of each other in a given line. If A' is the image of A in the line l, then one also says that *A' is symmetric to A with respect to the line l.*

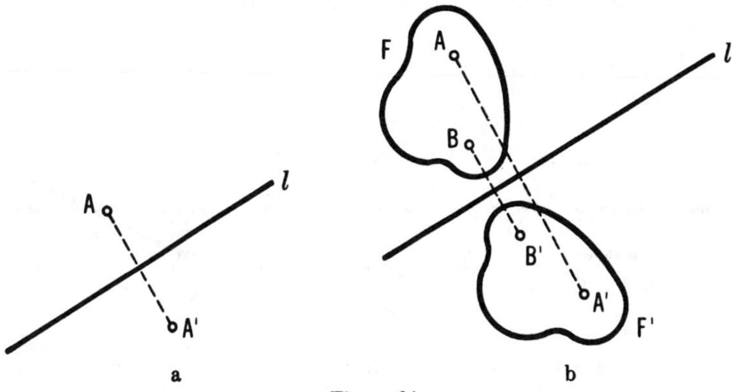

Figure 34

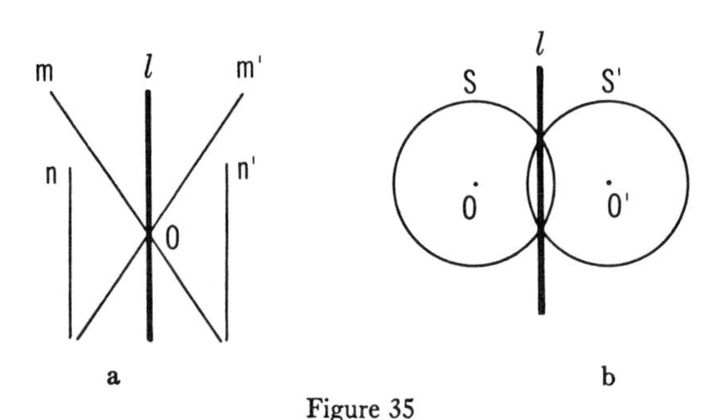

Figure 35

The set of all images in a line l of the points of a figure F forms a figure F', called the image of the figure F by reflection in l (Figure 34b); it is clear that, conversely, F is the image of F' in l. A line is taken by reflection in l into a new line; at the same time a line parallel to l is reflected into a line parallel to l, and a line meeting l in a point O is reflected into a line meeting l in the same point (in Figure 35a, n is taken into n', and m into m'). A circle is reflected into a congruent circle (Figure 35b). (To prove this last assertion, for example, it is sufficient to note that every segment AB is reflected into a segment $A'B'$ of the same length. Thus, in Figure 36a, $AB = PQ = A'B'$ and in Figures 36b, c, $AB = A'B'$ since $\Delta AOP \cong \Delta A'OP$, $\Delta BOQ \cong \Delta B'OQ$, and, therefore, $OA = OA'$, $OB = OB'$. From this it follows that the locus of points whose distance from O is equal to r is reflected into the locus of points whose distance from O' is equal to r, where O' is the reflection of O in the line, that is, the circle S is taken into the congruent circle S'.)

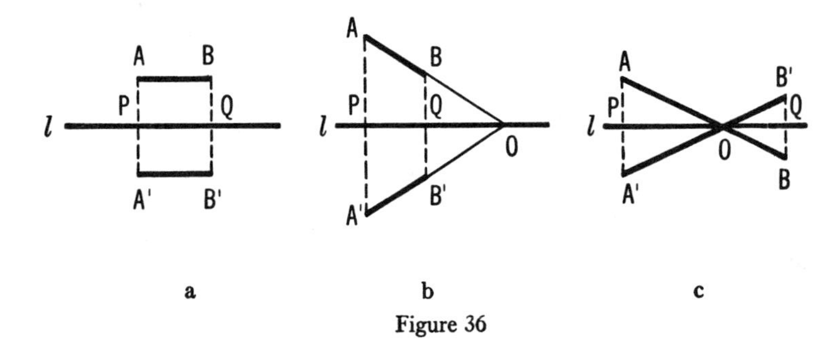

Figure 36

25. (a) Let a line MN be given together with two points A and B on one side of it. Find a point X on the line MN such that the segments AX and BX make equal angles with the line, i.e., such that

$$\angle AXM = \angle BXN.$$

(b) Let a line MN be given together with two circles S_1 and S_2 on one side of it. Find a point X on the line MN such that one of the tangents from this point to the first circle and one of the tangents from this point to the second circle make equal angles with the line MN.

(c) Let a line MN be given together with two points A and B on one side of it. Find a point X on the line MN such that the segments AX and BX make angles with this line, one of which is twice as large as the other (that is, $\angle AXM = 2\angle BXN$; see Figure 37).

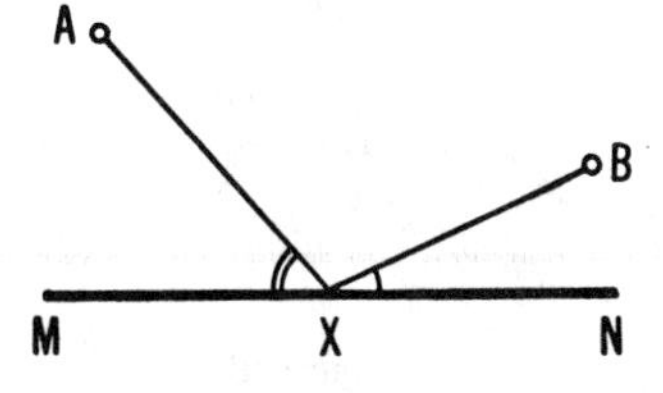

Figure 37

26. (a) Let lines l_1, l_2, and l_3, meeting in a point, be given, together with a point A on one of these lines. Construct a triangle ABC having the lines l_1, l_2, l_3, as angle bisectors.

(b) Let a circle S be given together with three lines l_1, l_2, and l_3 through its center. Find a triangle ABC whose vertices lie on the given lines, and such that the circle S is its inscribed circle.

(c) Let three lines l_1, l_2, l_3, meeting in a point, be given, together with the point A_1 on one of them. Find a triangle ABC for which the point A_1 is the midpoint of the side BC and the lines l_1, l_2, l_3 are the perpendicular bisectors of the sides of the triangle.

Problem 39(b) and (a) is a generalization of Problem 26(a) and (c).

27. (a) Construct a triangle, given the base $AB = a$, the length h of the altitude on this base, and the difference γ of the two angles at the base.

(b) Construct a triangle, given two sides and the difference γ of the angles they make with the third side.

28. Let an angle MON be given, together with two points A and B. Find a point X on the side OM such that the triangle XYZ, where Y and Z are the points of intersection of XA and XB with ON, is isosceles: $XY = XZ$ (Figure 38).

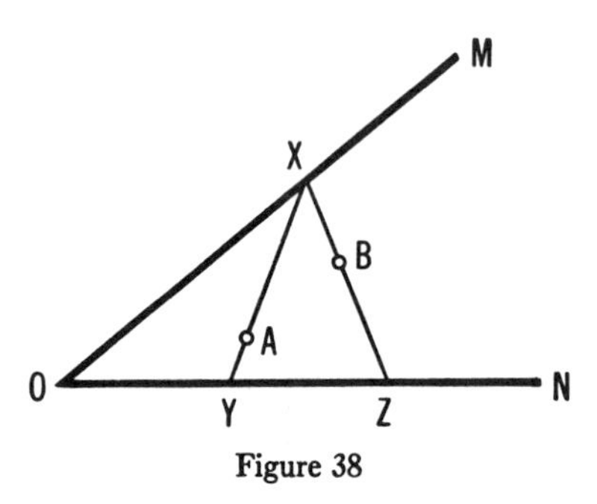

Figure 38

29. (a) Construct a quadrilateral $ABCD$ in which the diagonal AC bisects the angle A, given the lengths of the sides of the quadrilateral.

(b) Construct a quadrilateral in which a circle can be inscribed, given the lengths of two adjacent sides AB and AD together with the angles at the vertices B and D (the circle is to touch all four sides of the quadrilateral).

30. (a) A billiard ball bounces off a side of a billiard table in such a manner that the two lines along which it moves before and after hitting the sides are equally inclined to the side. Suppose a billiard table were bordered by n lines $l_1, l_2, \cdots, l_n$; let A and B be two given points on the billiard table. In what direction should one hit a ball placed at A so that it will bounce consecutively off the lines $l_1, l_2, \cdots, l_n$ and then pass through the point B (see Figure 39, where $n = 3$)?

(b) Let $n = 4$ and suppose that the lines l_1, l_2, l_3, l_4 form a rectangle and that the point B coincides with the point A. Prove that in this case the length of the total path of the billiard ball from the point A back to this point is equal to the sum of the diagonals of the rectangle (and, therefore, does not depend on the position of the point A). Prove also that if the ball is not stopped when it returns to the point A, then it will be reflected once more from the four sides of the rectangle and will return again to the point A.

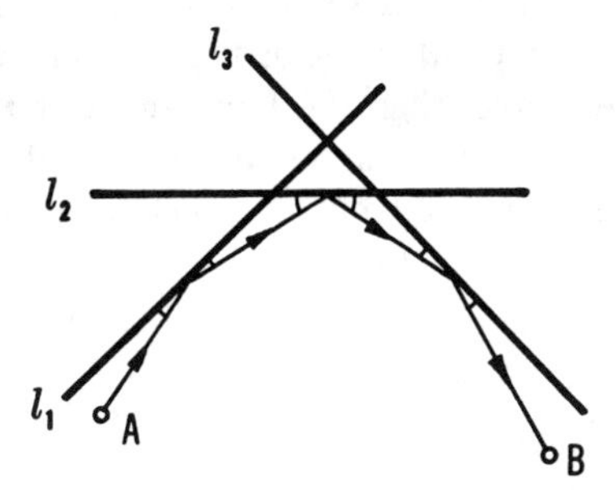

Figure 39

31\. (a) Let a line l and two points A and B on one side of it be given. Find a point X on the line l such that the sum $AX + XB$ of the distances has a given value a.

(b) Let a line l be given together with two points A and B on opposite sides of it. Find a point X on the line l such that the difference $AX - XB$ of the distances has a given value a.

32\. (a) Let ABC be any triangle and let H be the point of intersection of the three altitudes. Show that the images of H by reflection in the sides of the triangle lie on the circle circumscribed about the triangle.

(b) Given three points H_1, H_2, H_3 that are the images of the point of intersection of the altitudes of a triangle by reflection in the sides of the triangle; find the triangle.

The *orthocenter* of a triangle is the point of intersection of the three altitudes.

33. Let four points A_1, A_2, A_3, A_4 be given in the plane such that A_4 is the orthocenter of the triangle $A_1A_2A_3$. Denote the circles circumscribed about the triangles $A_1A_2A_3$, $A_1A_2A_4$, $A_1A_3A_4$, and $A_2A_3A_4$ by S_4, S_3, S_2, and S_1, and let the centers of these circles be O_4, O_3, O_2, and O_1. Prove that:

(a) A_1 is the orthocenter of triangle $A_2A_3A_4$, A_2 is the orthocenter of triangle $A_1A_3A_4$, and A_3 is the orthocenter of triangle $A_1A_2A_4$.

(b) The circles S_1, S_2, S_3, and S_4 are all congruent.

(c) The quadrilateral $O_1O_2O_3O_4$ is obtained from the quadrilateral $A_1A_2A_3A_4$ by means of a half turn about some point O (Figure 40). (In other words, if the points A_1, A_2, A_3, and A_4 are so placed that each point is the orthocenter of the triangle formed by the other three, then the four segments that connect each point to the center of the circle through the remaining three points all meet in one point O, which is the mid-point of each segment.)

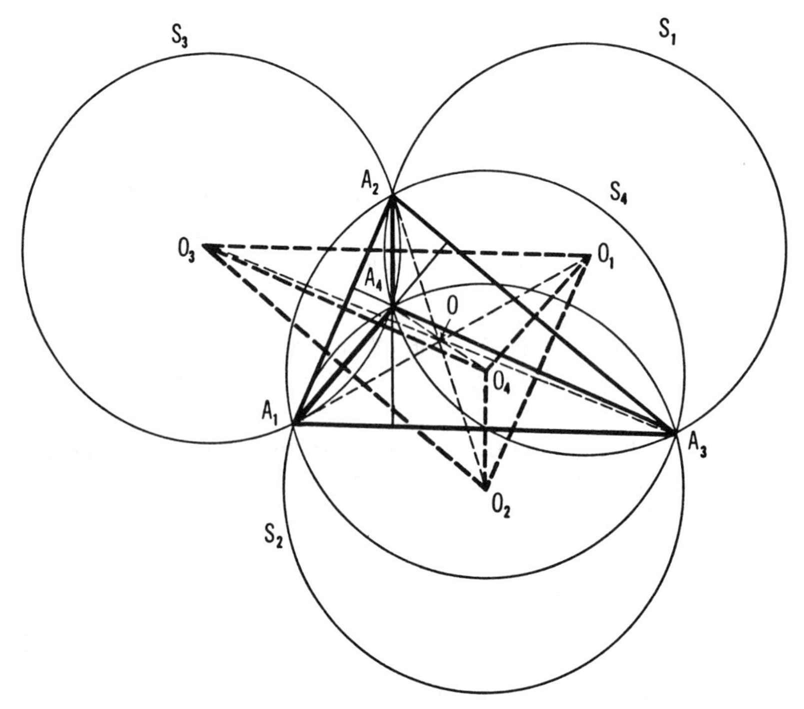

Figure 40

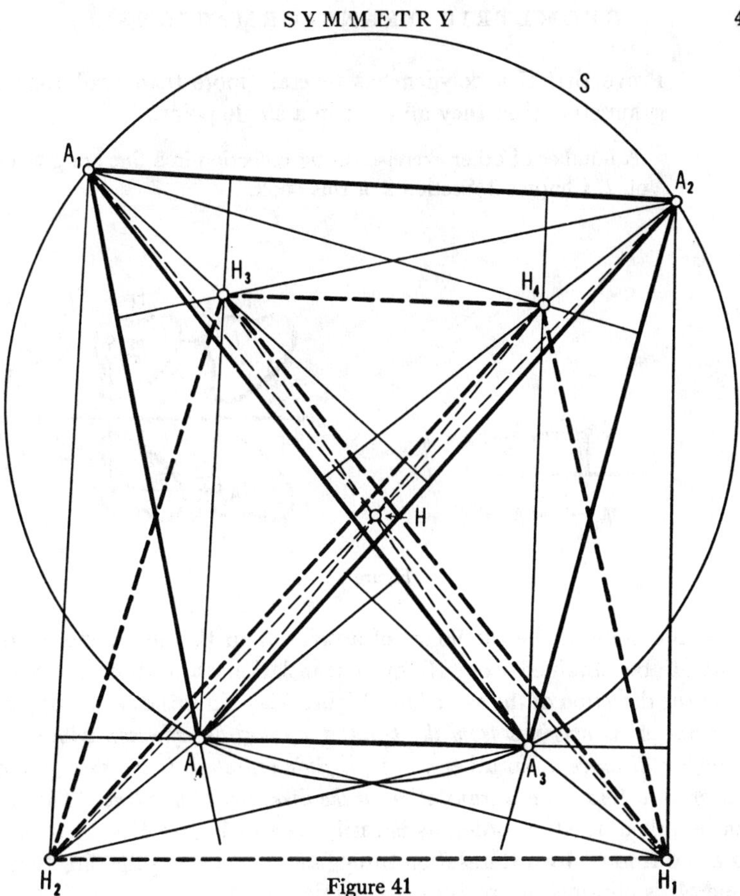

Figure 41

34. Let four points A_1, A_2, A_3, A_4 be given, all lying on a circle S. We denote the orthocenter of the triangles $A_1A_2A_3$, $A_1A_2A_4$, $A_1A_3A_4$, and $A_2A_3A_4$ by H_4, H_3, H_2, and H_1. Prove that:

(a) The quadrilateral $H_1H_2H_3H_4$ is obtained from the quadrilateral $A_1A_2A_3A_4$ by means of a half turn about some point H (Figure 41). (In other words, if the points A_1, A_2, A_3, A_4 all lie on one circle, then the four segments joining each of these points to the orthocenter of the triangle formed by the remaining three points meet in a single point, the midpoint of each segment.)

(b) The quadruples A_1, A_2, H_3, H_4; A_1, A_3, H_2, H_4; A_1, A_4, H_2, H_3; A_2, A_3, H_1, H_4; A_2, A_4, H_1, H_3; A_3, A_4, H_1, H_2; and H_1, H_2, H_3, H_4 each lie on a circle. Also, the seven circles on which these quadruples of points lie are all congruent to S.

35. Prove that if a polygon has several (more than two) axes of symmetry, then they all meet in a single point.

A number of other exercises using reflection in a line are given in Vol. 2, Chapter 2, Section 2 of this book.

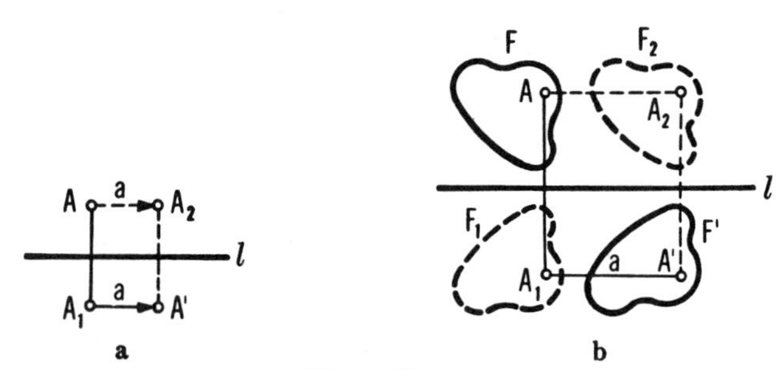

Figure 42

Let the point A_1 be the image of a point A in the line l, and let the point A' be obtained from A_1 by a translation through a distance a along the direction of the same line (Figure 42a). In this case we say that *the point A' is obtained from the point A by a glide reflection with axis l through a distance a.* In other words, *a glide reflection is the sum of a reflection in a line l and a translation in the direction of this line.* (The sum can be taken in either order, as is easily seen in Figure 42a; there A_2 is obtained from A by a translation through a distance a in the direction of l and A' is obtained from A_2 by a reflection in l.)

The set of all points that are obtained from the points of a figure F by means of a glide reflection forms a figure F' obtained by a glide reflection from the figure F (Figure 42b). It is clear that, conversely, the figure F can be obtained from F' by a glide reflection with the same axis l (and opposite direction of translation); this permits one to speak of figures related by a glide reflection.

36. Given a line l, two points A and B on one side of it, and a segment a; find a segment XY of length a on the line l, so that the length of the path $AXYB$ shall be as small as possible (Figure 43).

37. (a) Construct a quadrilateral $ABCD$ in which $\sphericalangle C = \sphericalangle D$, given the sides AB and CD, the sum of sides BC and AD, and the distance d from the vertex A to the side CD.

(b) Construct a quadrilateral $ABCD$, given the sides AB and CD, the sum of sides BC and AD, and the distances d_1 and d_2 from the vertices A and B to side CD.

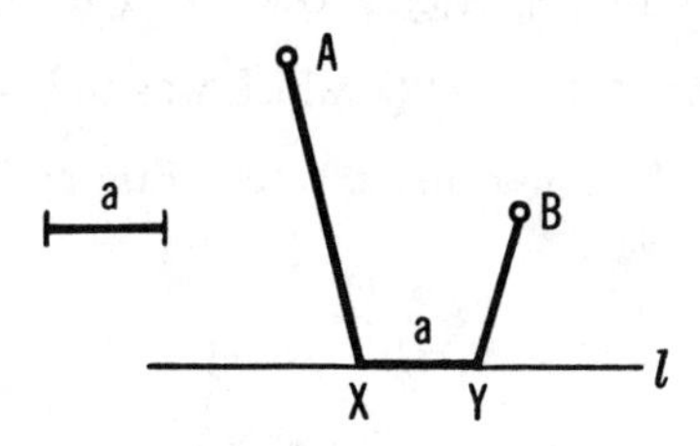

Figure 43

Now let us prove several propositions on the addition of reflections.[T]

PROPOSITION 1. *The sum of two reflections in one and the same line is the identity transformation.*

Indeed, if reflection in the line l carries the point A into the point A' (see Figure 34a), then a second reflection in l carries A' back into A; that is, as a result of two reflections the position of the point A is unchanged.

The assertion of Proposition 1 can also be formulated as follows: *Two reflections in the same line cancel each other.*

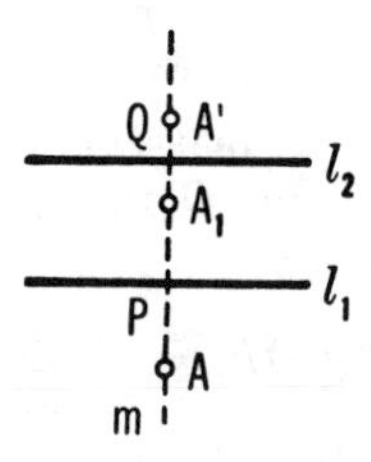

Figure 44a

PROPOSITION 2. *The sum of two reflections in parallel lines is a translation in the direction perpendicular to the two lines, through a distance equal to twice the distance between them.*

Let A be an arbitrary point in the plane, let A_1 be the reflection of A in the line l_1, and let A' be the reflection of A_1 in a line l_2 parallel to l_1 (Figure 44a). Then $AA_1 \perp l_1$ and $A_1A' \perp l_2$; consequently the points

[T] We shall frequently write *reflection* instead of *reflection in a line.*

A, A_1, and A' lie on a line m, perpendicular to l_1 and l_2. If P and Q are the points of intersection of the line m with l_1 and l_2, then $AP = PA_1$, $A_1Q = QA'$, and, for example, in the case pictured in Figure 44a,†

$$AA' = AP + PA_1 + A_1Q + QA' = 2PA_1 + 2A_1Q = 2PQ.$$

Thus, $AA' \perp l_1$ and $AA' = 2PQ$, which was to be proved.

Proposition 1 can be considered a special case of Proposition 2, namely the case when $PQ = 0$.

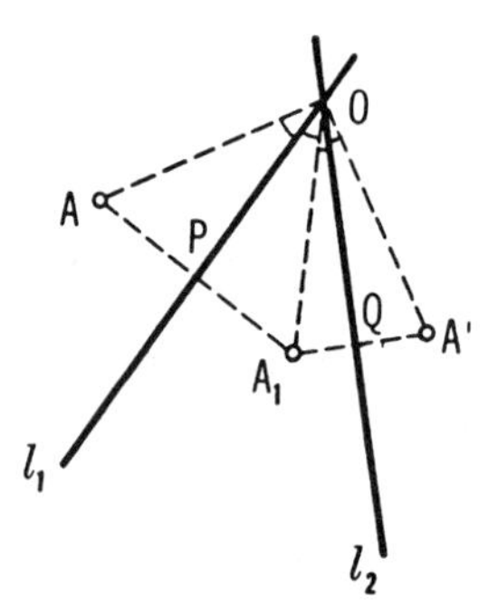

Figure 44b

PROPOSITION 3. *The sum of two reflections in intersecting lines is a rotation with center at the point of intersection of these lines, and through twice the angle between them.*

Let A be an arbitrary point of the plane, let A_1 be the image of A in the line l_1, and let A' be the image of A_1 in a line l_2 meeting l_1 in the point O (Figure 44b). If P and Q are the points of intersection of AA_1 with l_1 and of A_1A' with l_2, then

$$\Delta AOP \cong \Delta A_1OP, \qquad \Delta A_1OQ \cong \Delta A'OQ.$$

From this we have

$$OA = OA_1, \qquad OA_1 = OA';$$

$$\sphericalangle AOP = \sphericalangle POA_1, \qquad \sphericalangle A_1OQ = \sphericalangle QOA',$$

and, for example, in the case pictured in Figure 44b, ‡

† In order to carry out the proof without using the picture it is necessary to use the concept of directed line segment (see the small print on pages 20–21).

‡ To carry out this reasoning without dependence on a picture it is necessary to use the concept of directed angle (see the small print on page 30).

$$\sphericalangle AOA' = \sphericalangle AOP + \sphericalangle POA_1 + \sphericalangle A_1OQ + \sphericalangle QOA'$$
$$= 2\sphericalangle POA_1 + 2\sphericalangle A_1OQ$$
$$= 2\sphericalangle POQ.$$

Thus, $OA = OA'$ and $\sphericalangle AOA' = 2\sphericalangle POQ$, which was to be proved.†

Propositions 2 and 3 permit one to give a simple proof of the theorems on the addition of rotations or on the addition of a rotation and a translation.

Let it be required, for example, to find the sum of two rotations with centers O_1 and O_2 and angles α and β. By Proposition 3, the first rotation can be replaced by the sum of two reflections in lines l_1 and O_1O_2, where l_1 passes through O_1 and $\sphericalangle l_1O_1O_2 = \frac{1}{2}\alpha$; the second rotation can be replaced by the sum of two reflections in the lines O_1O_2 and l_2, where l_2 passes through O_2 and $\sphericalangle O_1O_2l_2 = \frac{1}{2}\beta$ (Figure 45). Thus the sum of two rotations is replaced by the sum of four reflections in the lines l_1, O_1O_2, O_1O_2, and l_2. But the middle two of these four reflections have the same axis and thus by Proposition 1 they cancel each other. Thus the sum of the four reflections in the lines l_1, O_1O_2, O_1O_2, and l_2 is identical with the sum of the two reflections in the lines l_1 and l_2. If O is the point of intersection of l_1 and l_2, then by Proposition 2 the sum of these two reflections is a rotation with center O and angle $2\sphericalangle l_1OO_2$, which, as one sees from Figure 45a, is equal to the sum of the angles

$$2\sphericalangle l_1O_1O_2 = \alpha \qquad \text{and} \qquad 2\sphericalangle O_1O_2l_2 = \beta$$

($\sphericalangle l_1OO_2$ is an exterior angle of the triangle O_1O_2O).

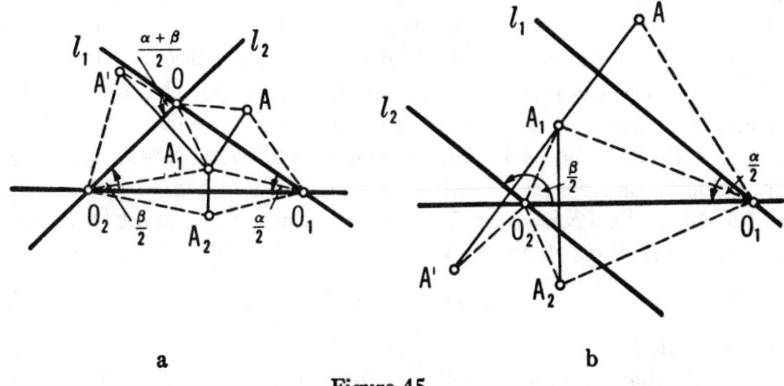

Figure 45

† From the proofs of Propositions 2 and 3 it is not difficult to see that the sum of two reflections in lines depends on the order in which these reflections are carried out (with the exception of the one case when the lines are perpendicular and the sum of the reflections taken in either order is a half turn about their point of intersection).

If l_1 and l_2 are parallel (from Figure 45b, it is clear that this case will occur when $\sphericalangle l_1O_1O_2 + \sphericalangle O_1O_2l_2 = 180°$, that is, when $\alpha + \beta = 360°$), then by Proposition 2 the sum of the reflections in l_1 and l_2 is a translation. Thus we again come to the same result as before (see page 34).

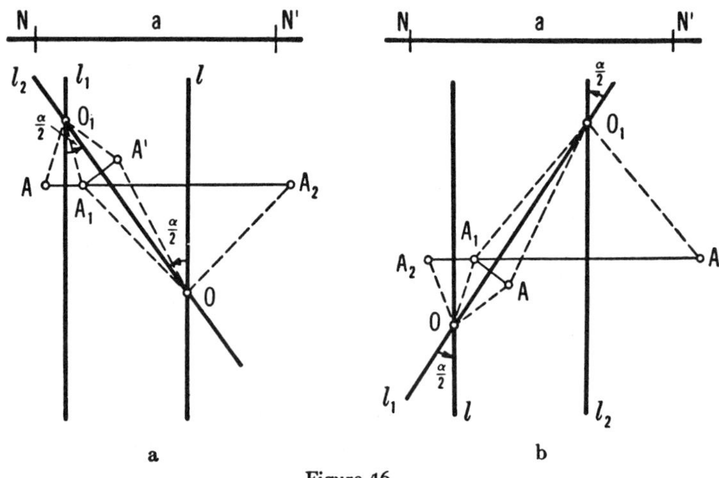

Figure 46

Now let us find the sum of a translation in the direction NN' through a distance a and a rotation with center O and angle α. We replace the translation by the sum of two reflections in lines l_1 and l, perpendicular to NN', so that the distance between them is $\frac{1}{2}a$, and we let l pass through O (Figure 46a). We replace the rotation by the sum of two reflections in the lines l and l_2, where l_2 passes through O and $\sphericalangle lOl_2 = \frac{1}{2}\alpha$. Thus the sum of a translation and a rotation is replaced by the sum of four reflections in the lines l_1, l, l, and l_2. The middle two of these four reflections cancel each other by Proposition 1; thus we are left with the sum of the two reflections in the lines l_1 and l_2, which by Proposition 3 is a rotation about the point O_1 of intersection of l_1 and l_2, through an angle

$$2\sphericalangle l_1O_1l_2 = 2\sphericalangle lOl_2 = \alpha$$

(see Figure 46a).

In exactly the same way it is shown that the sum of a rotation with center O and angle α and a translation in the direction NN' through a distance a is a rotation with the same angle of rotation α. To find the center O_1 of this rotation, one passes lines l and l_1 through O, with $l \perp NN'$ and $\sphericalangle l_1Ol = \frac{1}{2}\alpha$, and a line $l_2 \parallel l$ at a distance of $\frac{1}{2}a$ from l. O_1 is then the point of intersection of l_1 and l_2 (Figure 46b).

PROPOSITION 4. *The sum of the reflections in three parallel lines or in three lines meeting in a single point is a reflection in a line.*

Let us assume first that the three lines l_1, l_2, and l_3 are parallel (Figure 47a). By Proposition 2 the sum of the reflections in the lines l_1 and l_2 is a translation in the direction perpendicular to l_1 and l_2 through a distance equal to twice the distance between them, and coincides with the sum of the reflections in any other two lines l and l' that are parallel to l_1 and l_2 and the same distance apart. Now assume that l' coincides with l_3, and replace the sum of our three reflections by the sum of the reflections in the lines l, l', and l_3. By Proposition 1, the last two of these reflections cancel each other and so there remains only the reflection in the line l.

Now let the lines l_1, l_2, and l_3 meet in a point O (Figure 47b). By Proposition 3 the sum of the reflections in l_1 and l_2 is a rotation about O through an angle $2\sphericalangle l_1Ol_2$ and coincides with the sum of the reflections in the lines l and l_3, where l passes through O and $\sphericalangle lOl_3 = \sphericalangle l_1Ol_2$. Therefore the sum of the reflections in l_1, l_2, and l_3 is equal to the sum of the reflections in l, l_3, and l_3 or to a single reflection in l (because the last two reflections in l_3 cancel each other).

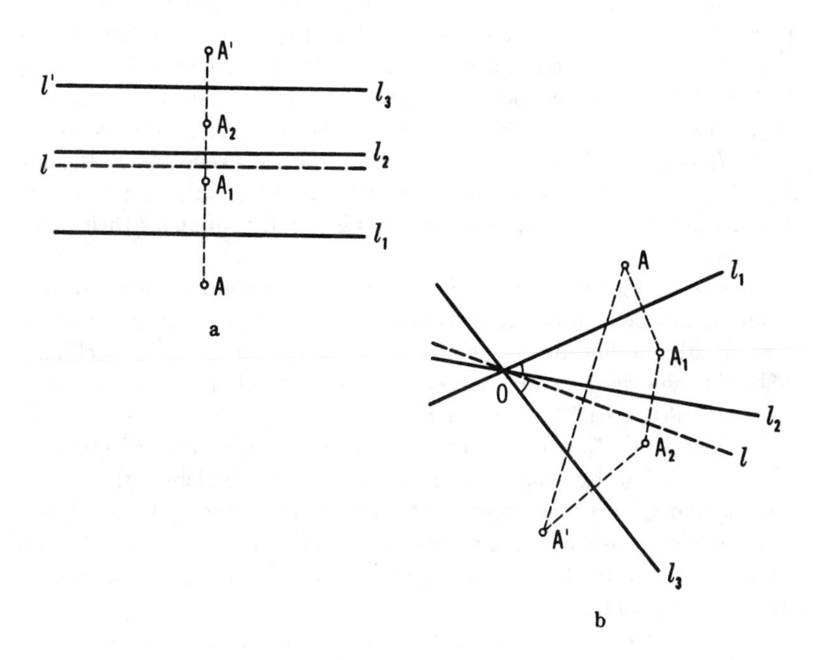

Figure 47

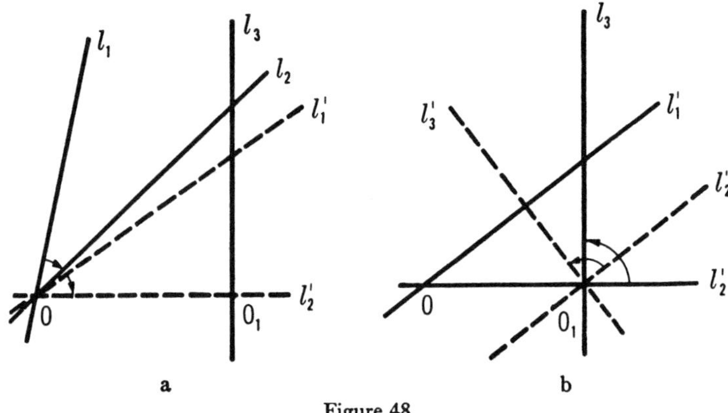

Figure 48

PROPOSITION 5. *The sum of the reflections in three lines, intersecting in pairs in three points, or such that two of them are parallel and the third intersects them, is a glide reflection.*

Let the lines l_1 and l_2 meet in the point O (Figure 48a). The sum of the reflections in l_1 and l_2 is a rotation with center O and angle $2\measuredangle l_1Ol_2$ (see Proposition 3); therefore the sum of these reflections can be replaced by the sum of the reflections in any two lines l_1' and l_2', meeting in the same point O and forming the same angle as l_1 and l_2. Choose the lines l_1' and l_2' such that $l_2' \perp l_3$, and replace the sum of the reflections in l_1, l_2, and l_3 by the sum of the reflections in the lines l_1', l_2', and l_3 (that is, by the sum of a reflection in l_1' and a half turn about the point O_1 of intersection of l_2' and l_3—or, what is the same thing, by the sum of a reflection in the line l_1' and a reflection in the point O_1—because by Proposition 3 the sum of the reflections in two perpendicular lines is a half turn about their point of intersection).

Now let us replace the sum of the reflections in the perpendicular lines l_2' and l_3 by the sum of the reflections in two new perpendicular lines l_2'' and l_3', intersecting in the same point O_1, and such that $l_2'' \parallel l_1'$ (Figure 48b; this change is permissible because the sum of the reflections in l_2'' and l_3' is also a half turn about O_1). At the same time the sum of the reflections in l_1', l_2', and l_3 is replaced by the sum of the reflections in l_1', l_2'', and l_3'. But by Proposition 2 the sum of the reflections in the parallel lines l_1' and l_2'' is a translation in the direction l_3' perpendicular to l_1' and l_2''. Therefore the sum of the reflections in l_1', l_2'', and l_3' is equal to the sum of a translation in the direction l_3' and a reflection in l_3', that is, a glide reflection with axis l_3'.

In case l_1 and l_2 are parallel, and l_2 and l_3 intersect in a point O, the proof proceeds in exactly the same way. (In this case it is necessary first to replace the sum of the reflections in l_2 and l_3 by the sum of the reflections in lines l_2' and l_3' intersecting in the same point O, and such that $l_2' \perp l_1$; and then to replace the sum of the reflections in the perpendicular lines l_1 and l_2' by the sum of the reflections in the perpendicular lines l_1' and l_2'', intersecting in the same point O_1 and such that $l_2'' = l_3'$.)

From Propositions 2–5 we obtain the following general

THEOREM. *The sum of an even number of reflections is a rotation or a translation; the sum of an odd number of reflections is a reflection or a glide reflection.*

Indeed, the sum of an even number of reflections may, by Propositions 2 and 3, be replaced by the sum of a number of rotations and translations. But the sum of any number of rotations and translations is again a rotation or a translation (on this see Chapter 1 or the text in small print on pages 51–52).

Further, since the sum of an even number of reflections is a rotation or a translation, the sum of an odd number of reflections may be replaced by the sum of a rotation or a translation and a reflection. By Propositions 2 and 3, a rotation or translation can be replaced by the sum of two reflections. Thus the sum of an odd number of reflections can always be replaced by a sum of three reflections, and these may be treated by Propositions 4 and 5.

Let us note also that the sum of an even number of reflections is, generally speaking, a rotation; the cases when this sum reduces to a translation must be regarded as exceptional. (The sum of two reflections in the lines l_1 and l_2 is a translation only in the exceptional case where $l_1 \parallel l_2$; the sum of two rotations through the angles α and β is a translation only in the exceptional case where $\alpha + \beta = 360°$, and so forth.) Analogously, the sum of an odd number of reflections is, generally speaking, a glide reflection; the cases in which the sum of an odd number of reflections reduces to a reflection must be considered as exceptional. (For example, the sum of three reflections in the lines l_1, l_2, and l_3 is a reflection only in the exceptional cases when the lines l_1, l_2, and l_3 are all parallel or all intersect in a single point.)

Reflection and glide reflection are transformations of the plane carrying each point A into a new point A'.† *The fixed points of a reflection in l*

† Reflection is an isometry in the sense of the definition given in the introduction to this part, because this transformation carries each segment AB into a segment $A'B'$ of the same length (see Figure 36 and the accompanying text). Glide reflection is an isometry because it is the sum of two isometries: reflection and translation.

are the points of the axis of symmetry l; the fixed lines are the axis l and all lines perpendicular to l. The only fixed line of a glide reflection is its axis l; glide reflection has no fixed points whatsoever.

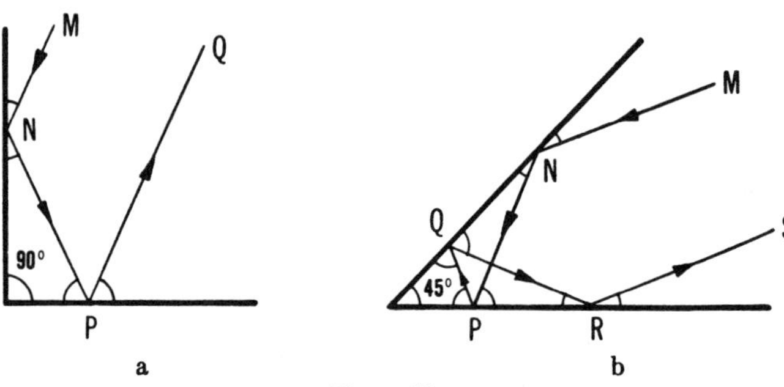

Figure 49

38. A ray of light is reflected from a straight-line mirror in such a way that the angle of incidence is equal to the angle of reflection (that is, by the same law as that by which a billiard ball rebounds from the side of a billiard table; see Problem 30). Let two straight-line mirrors be given in the plane, forming an angle α. Prove that if $\alpha = 90°/n$, where n is a whole number (and only in this case!), then any light ray, after being reflected several times in both mirrors will, in the end, pass off in a direction exactly opposite to the direction of first approach [see Figures 49a, b, where the cases $n = 1$, $\alpha = 90°$ and $n = 2$, $\alpha = 45°$ are shown; in both cases the final direction of the ray (PQ and RS respectively) is opposite to the original direction MN].

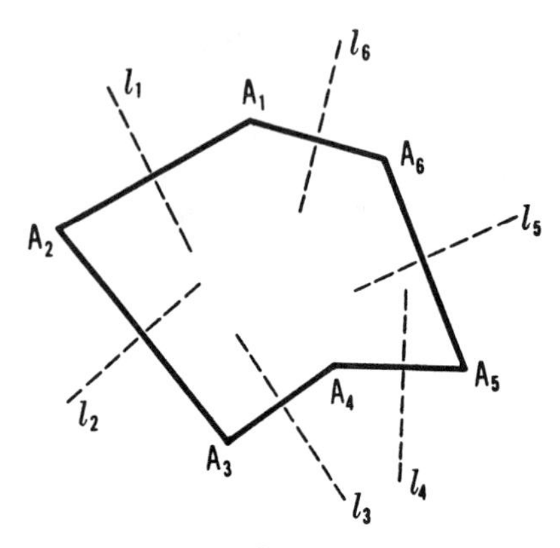

Figure 50a

39. Let n lines $l_1, l_2, \cdots, l_n$ be given in the plane. Construct an n-gon $A_1A_2 \cdots A_n$ for which these lines are:

(a) The perpendicular bisectors of its sides (Figure 50a).

(b) The bisectors of the exterior or of the interior angles at the vertices (Figure 50b).

Consider separately the cases of odd and even n. In which case will the problem have no solution, or will the solution not be uniquely determined?

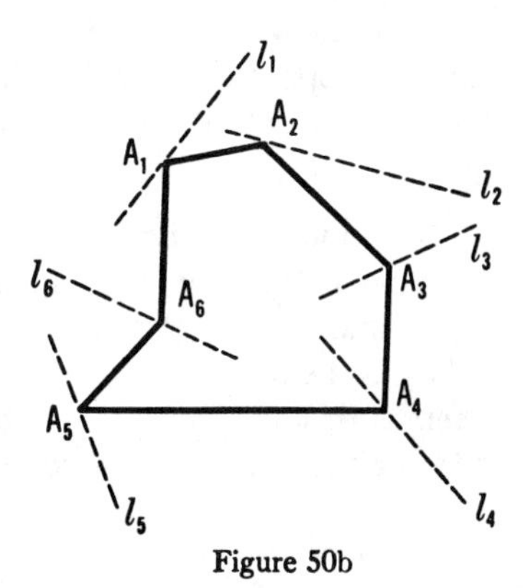

Figure 50b

40. Let a point M and $n-1$ lines $l_2, l_3, \cdots, l_n$ be given in the plane. Construct an n-gon $A_1A_2 \cdots A_n$:

(a) Such that the midpoint of side A_1A_2 coincides with the point M, and the perpendicular bisectors of the remaining sides coincide with the lines $l_2, l_3, \cdots, l_n$.

(b) Such that the angle A_1 has a given value α, its bisector passes through M, and the bisectors of the angles $A_2, A_3, \cdots, A_n$ coincide with $l_2, l_3, \cdots, l_n$.

41. In a given circle inscribe an n-gon:

(a) The sides of which are parallel to n given lines in the plane.

(b) The side A_1A_n of which passes through a given point, and the remaining sides are parallel to $n-1$ given lines.

42. (a) Let three lines l_1, l_2, and l_3 meeting in a point be given in the plane. Let an arbitrary point A in the plane be reflected successively in l_1, l_2, and in l_3; then let the point A_3 thus obtained be reflected successively again in these same three lines in the same order. Show that the final point A_6 obtained as a result of these six reflections coincides with the original point A (Figure 51a).

Does the conclusion of this problem remain valid if the three lines l_1, l_2, l_3 are replaced by n arbitrary lines in the plane, all meeting in a common point (the six reflections now become $2n$ reflections)?

(b) Let three lines l_1, l_2, and l_3 meeting in a point be given in the plane. An arbitrary point A in the plane is reflected successively in l_1, l_2, and in l_3; then the same point A is reflected in the same three lines taken in the opposite order, first in l_3, then in l_2, and finally in l_1. Show that in both cases we are led to one and the same final point A_3.

(c) Let four lines l_1, l_2, l_3, and l_4 meeting in a point be given in the plane. An arbitrary point A in the plane is reflected successively in the lines l_1, l_2, l_3, and l_4; then this same point A is reflected successively in these same lines but in a different order: first in l_3, then in l_4, then in l_1, and, finally, in l_2. Show that in both cases we are led to one and the same final point A_4 (Figure 51b).

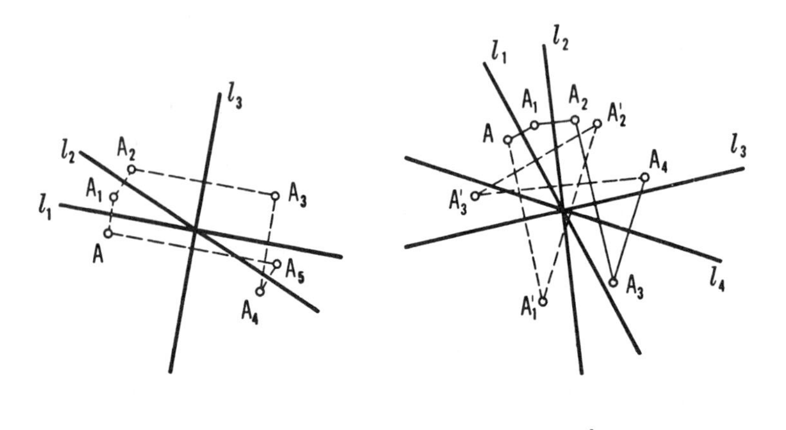

Figure 51

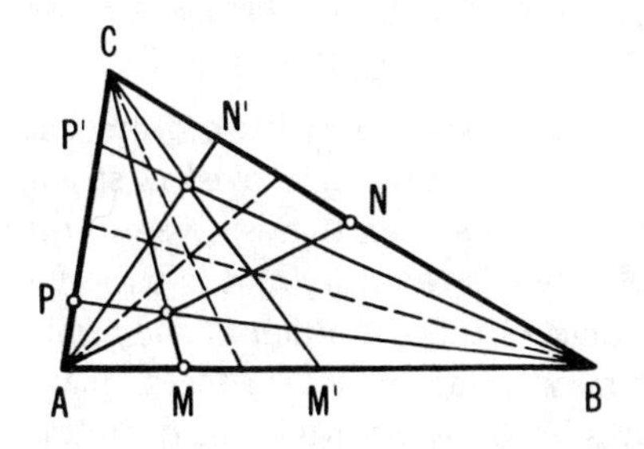

a

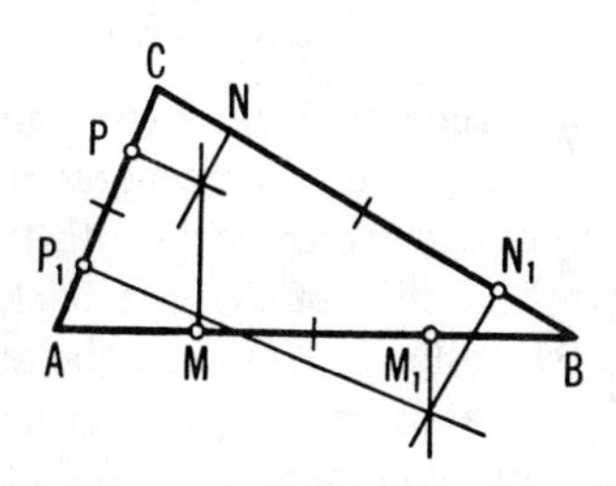

b

Figure 52

43. (a) Let M, N, P be points on the sides AB, BC, and CA of triangle ABC. Let CM', AN', BP' be the images of CM, AN, and BP in the bisectors of the angles C, A, B, respectively, of the triangle. Show that if the lines CM, AN and BP meet in a point or are all parallel to one another, then the lines CM', AN', and BP' also meet in a point or are all parallel to one another (Figure 52a).

(b) Let M, N, P be points on the sides AB, BC, and CA of triangle ABC, and let M_1, N_1, P_1 be the images of M, N, and P in the midpoints of the corresponding sides of the triangle (that is, M_1 is obtained from M by a half turn about the midpoint of side AB, and similarly for the other points). Show that if the perpendiculars to AB, BC, and CA, erected at the points M, N, and P meet in a point, then the perpendiculars to AB, BC, and CA, erected at the points M_1, N_1, and P_1 also meet in a point (Figure 52b).

44. Let three arbitrary lines l_1, l_2, and l_3 be given in the plane. An arbitrary point A of the plane is reflected twice in these three lines: in l_1, l_2, l_3 and again in l_1, l_2, l_3; the point A_6 obtained from A as a result of these six reflections is now reflected in these same lines but in a different order: in l_2, l_3, l_1 and again in l_2, l_3, l_1. Now start over again, this time reflecting the original point A successively in l_2, l_3, l_1 and again in l_2, l_3, l_1; the point A_6' obtained from these six reflections is now reflected twice in the three lines l_1, l_2, l_3, taken in this order. Show that in each case at the end of the twelve reflections we come to one and the same point A_{12}.

2. Directly Congruent and Oppositely Congruent Figures. Classification of Isometries of the Plane

According to Kiselyov's high school geometry text, "two geometric figures are said to be congruent if one figure, by being moved in space, can be made to coincide with the other." This definition is given in the very beginning of the first book of Kiselyov's *Geometry* and is basic for all that follows. However, the appearance of this definition at the beginning of a course in plane geometry may arouse doubts. Indeed, plane geometry considers properties of figures in the *plane*, while the definition of congruence speaks of moving figures in *space*. Thus it would seem that the first and most basic definition in a course in plane geometry does not itself belong to plane geometry at all, but to solid geometry. It would seem to be more proper in a course in plane geometry to say that two figures will be called congruent if they can be made to coincide by moving them in the plane, and not in space—such a definition would not use concepts of solid geometry. But it turns out that this new definition of congruence is not entirely equivalent to the original one.

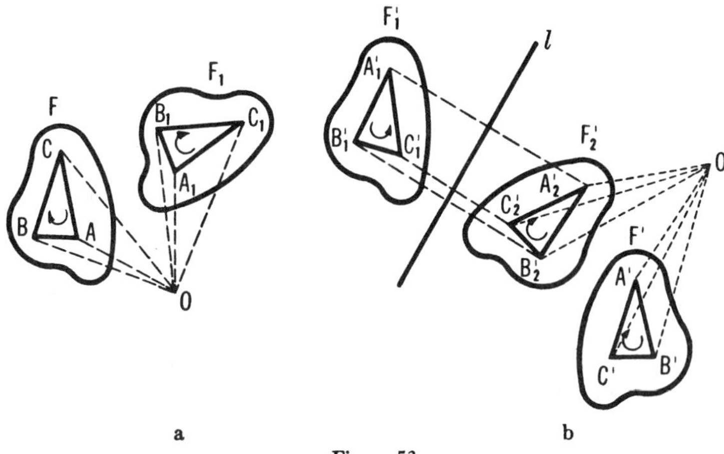

Figure 53

Indeed, pairs of congruent figures in the plane can be of two types. It is possible that two congruent figures can be made to coincide by moving one of the figures, but without removing it from the plane in which it was originally situated; such, for example, are the figures F and F_1 in Figure 53a (they can be made to coincide by means of a rotation about the point O). But it is also possible for two plane figures to be congruent although, to make them coincide, it is necessary to remove one of them from the plane and to turn it over "onto its other side". Such are figures F' and

F'_1 in Figure 53b; it is impossible to move the figure F'_1 in the plane so that it coincides with the figure F'.

Indeed, let us consider three points A'_1, B'_1, and C'_1 of the figure F'_1 and the corresponding points A', B', and C' of the figure F'. The triangles $A'B'C'$ and $A'_1B'_1C'_1$ have, as is said, "different orientations": In triangle $A'B'C'$ the direction around the boundary from vertex A' to vertex B' to vertex C' coincides with the direction in which the hands of a clock move (clockwise direction), while in triangle $A'_1B'_1C'_1$ the direction around the boundary from vertex A'_1 to vertex B'_1 to vertex C'_1 is opposite to the direction in which the hands of a clock move (counterclockwise direction). And since obviously in any movement entirely in the plane of the figure F'_1, the orientation of triangle $A'_1B'_1C'_1$ cannot change, we cannot possibly make triangle $A'_1B'_1C'_1$ coincide with triangle $A'B'C'$. But if we "turn figure F'_1 over onto its other side"—for this it is sufficient to replace F'_1 by the figure F'_2 obtained by reflecting F'_1 in some line l—then there is no difficulty in moving F'_2 in the plane so that it coincides with F' (a rotation about the point O; see Figure 53b).

In what follows figures that can be made to coincide by a motion entirely within the plane will be called *directly congruent*; congruent figures that cannot be made to coincide by motions entirely within the plane will be called *oppositely congruent*. From what has already been said it is clear how to decide whether two given congruent figures F and F' are directly congruent or oppositely congruent: It is sufficient to choose any three points A, B, C of the figure F and the corresponding points A', B', C' of the figure F', and to check whether the orientations of the triangles ABC and $A'B'C'$ (from A to B to C, respectively from A' to B' to C') are the same or are opposite. We shall call two figures "congruent" only in case it does not matter to us whether they are directly congruent or oppositely congruent.

Thus, *two geometric figures will be called directly congruent if one of them can be moved entirely in the plane so that it coincides with the second figure.* This definition is almost word for word the same as Kiselyov's definition of congruence, but it belongs entirely to plane geometry.

We now prove two important theorems.

THEOREM 1. *Any two directly congruent figures in the plane can be made to coincide by means of a rotation or a translation.*

Let us first note that any two congruent line segments AB and $A'B'$ in the plane can be made to coincide by a rotation or by a translation.

Indeed, if the segments AB and $A'B'$ are equal, parallel, and have the same direction (Figure 54a) then AB can be carried into $A'B'$ by a translation (see pages 18–19 where a more general proposition was proved about two figures F and F', corresponding segments of which are equal, parallel, and have the same direction); the distance and direction of this translation are determined by the segment AA'. If the segments AB and $A'B'$ make an angle α (Figure 54b),† then AB can be carried into $A'B'$ by a rotation through an angle α (see page 32 where a more general proposition is proved about two figures F and F', corresponding segments of which are equal and make an angle α); the center O of this rotation can be found, for example, as the point of intersection of the perpendicular bisectors of the segments AA' and BB'.‡

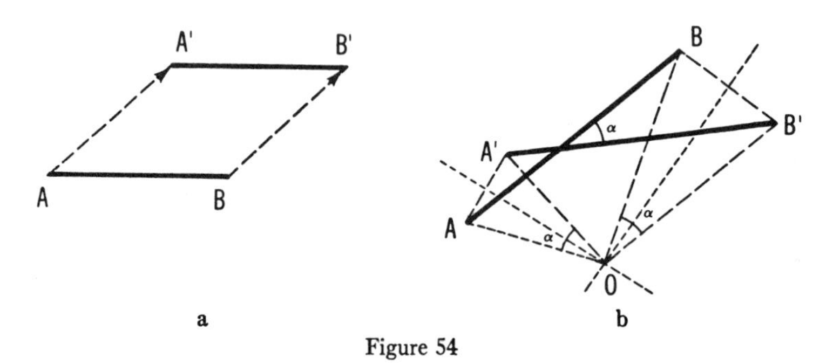

Figure 54

Now let us consider two directly congruent figures F and F' (Figure 55). Let M and N be any two points of the figure F; points M' and N' of the figure F' correspond to them. Since the figures are congruent, $MN = M'N'$ and, therefore, there exists a rotation (or a translation) carrying the segment MN into the segment $M'N'$.

† This includes the case when the segments AB and $A'B'$ make an angle $\alpha = 180°$, that is, when they are equal, parallel, and oppositely directed.

‡ If these perpendiculars coincide then this construction doesn't work; in this case O is the point of intersection of the segments AB and $A'B'$ themselves (and if these segments coincide, that is, if A coincides with B' and B with A', then O is the common midpoint of AB and $A'B'$). O can also be found as the point of intersection of the circular arc constructed on AA' and subtending an angle α with it, and the perpendicular bisector of AA'. Finally two more convenient constructions of the center of rotation carrying a given segment AB into another given segment $A'B'$ are given in Vol. 2, Chapter 1, Section 2.

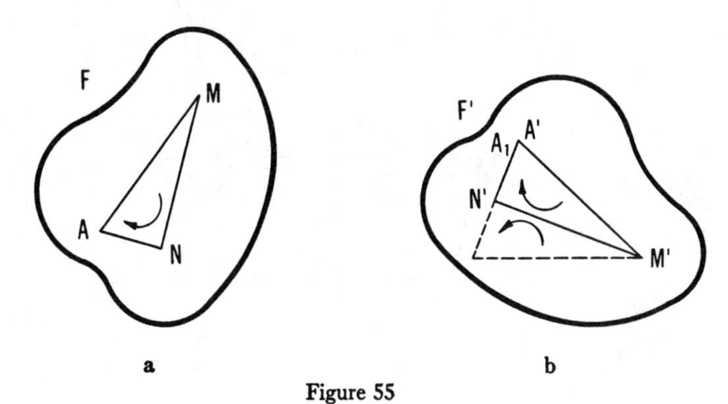

Figure 55

We assert that actually the whole figure F is carried into the figure F', that is, that each point A of the figure F is carried into its corresponding point A' of the figure F'. Denote by A_1 the point into which the point A is taken by the rotation (or translation) carrying MN into $M'N'$; we must prove that A_1 coincides with A'. Since the figures F and F' are congruent, $AM = A'M'$, $AN = A'N'$; on the other hand, it is clear that $AM = A_1M'$, $AN = A_1N'$. From this it follows that the triangles $A'M'N'$ and $A_1M'N'$ are congruent. And since these triangles have a common side $M'N'$, they must either coincide or be images of each other in the line $M'N'$. It only remains to show that this last case is impossible. The triangles AMN and $A'M'N'$ have the same orientation because the figures F and F' are directly congruent; the triangles AMN and $A_1M'N'$ also have the same orientation, since they are related by a rotation or translation. Hence the triangles $A'M'N'$ and $A_1M'N'$ have the same orientation and therefore cannot be oppositely congruent. This means that they coincide, and the point A is indeed carried by the rotation (or translation) into the point A'. This completes the proof of Theorem 1.

If the figures F and F' can be carried into each other by a rotation with center O, then the point O is called the *center of rotation* of these two figures. To construct the center of rotation O of two directly congruent figures, it is sufficient to choose any two points A and B of one figure and the corresponding points A' and B' of the second figure; O is the point of intersection of the perpendicular bisectors of AA' and BB' (see the footnote ‡ on page 62).

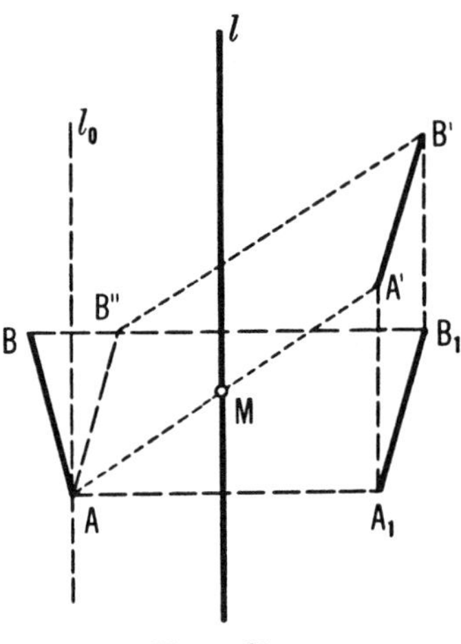

Figure 56

THEOREM 2. *Any two oppositely congruent figures in the plane can be made to coincide by means of a glide reflection or a reflection.*

The proof of Theorem 2 is analogous to the proof of Theorem 1. First of all we show that any two equal line segments AB and $A'B'$ can be carried into each other by means of a glide reflection with some axis l (or by means of a reflection in some line l). Indeed, assume that this is true and let l be the axis of the glide reflection (or of the reflection). Translate the segment $A'B'$ into a new position $A''B''$ so that A' is carried into A (i.e., $A'' = A$; see Figure 56). Since the segment A_1B_1 obtained from AB by reflection in the line l must be parallel to $A''B''$ (both these segments are parallel to $A'B'$), it follows that the line l must be parallel to the bisector l_0 of the angle $B''AB$ (because the sum of the reflections in l_0 and l carries the segment $A''B''$ into the parallel segment A_1B_1). Further, the points A and A' must be placed at equal distances from the line l and on different sides of it (because the points A and A_1 are placed at equal distances from l on opposite sides of it, and the points A_1 and A' are placed at equal distances from l on the same side of it). From this it follows that the line l must pass through the midpoint M of the segment AA'. Thus, if we know the segments AB and $A'B'$, then we can construct the line l (it is parallel to l_0 and passes through M).

Now let the segment A_1B_1 be the image of AB in the line l. Since $l \parallel l_0$, we have $A_1B_1 \parallel A'B'$; since l passes through M, it follows that the points A_1 and A' are equidistant from l and on the same side of it. Consequently, if the segment A_1B_1 does not coincide with $A'B'$, then it can be carried into $A'B'$ by a translation in the direction of the line l. From this it is clear that the segment AB can indeed be carried into the equal segment $A'B'$ by a glide reflection (or by a reflection).

The concluding part of the proof of Theorem 2 is almost an exact repetition of the last part of the proof of Theorem 1. Let F and F' be any two oppositely congruent figures and M, N and M', N' any two pairs of corresponding points of these figures (Figure 57). There exists a glide reflection (or a reflection) carrying the segment MN into the segment $M'N'$. Let us show that actually the whole figure F is carried into the figure F' by this glide reflection (or reflection), that is, that the point A_1 into which a given point A of the figure F is taken by the glide reflection (or reflection) coincides with the point A' of F' corresponding to the point A of F (F and F' are oppositely congruent, and therefore to each point A of F there is a corresponding point A' of F'). Indeed, $\Delta A'M'N' \cong \Delta AMN$ since the figures F and F' are congruent;

$$\Delta A_1M'N' \cong \Delta AMN$$

since $A_1M'N'$ is obtained from AMN by a glide reflection (or by a reflection). Therefore $\Delta A_1M'N'$ either coincides with $\Delta A'M'N'$ or is the image of $\Delta A'M'N'$ in the common side $M'N'$ of these two triangles. But the triangles $A_1M'N'$ and $A'M'N'$ cannot be images of each other, since they have the same orientation. This follows from the fact that the orientations of the triangles $A'M'N'$ and AMN are opposite (because the figures are oppositely congruent); the orientations of the triangles $A_1M'N'$ and AMN are also opposite (because reflection and glide reflection reverse the orientation of a triangle). Therefore the triangle $A_1M'N'$ must coincide with the triangle $A'M'N'$, which completes the proof of Theorem 2.

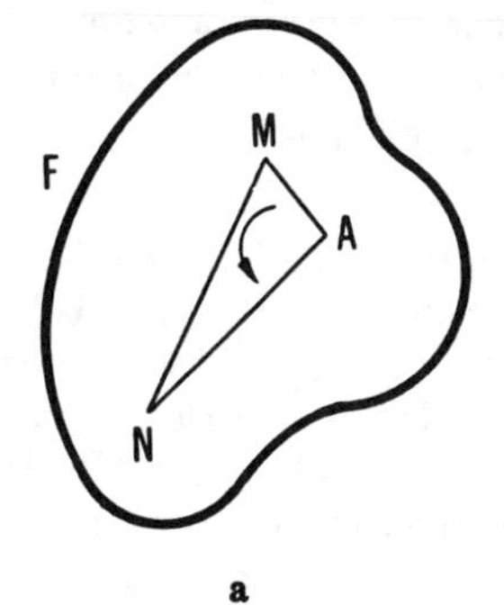

a

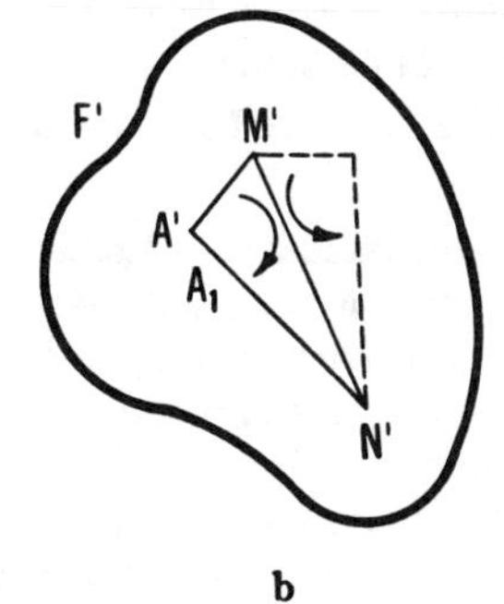

b

Figure 57

The isometries that carry directly congruent figures into each other are often called *direct isometries* (or *displacements*); in contrast to this, the isometries that carry two oppositely congruent figures into each other are called *opposite isometries*. Theorems 1 and 2 assert that each direct isometry is either a translation or a rotation while each opposite isometry is either a reflection or a glide reflection (compare this with the text on pages 68–69).

Combining the results of Theorems 1 and 2, one can formulate the following general assertion:

Any two congruent figures in the plane can be brought into coincidence by means of a translation or a rotation or a reflection or a glide reflection.

At the same time, if two figures are directly congruent, then in general they can be made to correspond by means of a rotation; the cases when the figures are related by a translation can be considered exceptional. If the figures are oppositely congruent, then they will, in general, be related by a glide reflection; the cases when the figures are related by a reflection can be considered exceptional.

Translation and rotation can be represented as the sum of the reflections in two (parallel or intersecting) lines, while reflection in a line or glide reflection can be represented as the sum of a reflection in a line and in a point (reflection in a line m is equal to the sum of the reflections in three lines: $l \perp m$, l, and m, that is, to the sum of the reflections in l and in the point O of intersection of l and m; as regards glide reflection, see page 48). Therefore our result can also be formulated as follows:

Any two congruent figures in the plane can be brought into coincidence by the sum of the reflections in two lines l_1 and l_2 or of the reflections in a line l and a point O. When $l_1 \parallel l_2$ we have a translation, and when the point O lies on the line l we have a reflection in a single line.

Theorems 1 and 2 can also be derived from the propositions on the addition of reflections (see pages 49–54). Indeed, the proof of Theorem 1 is based on the fact that each two equal line segments AB and $A'B'$ can be made to coincide by means of a rotation or by a translation. But clearly AB can be carried into $A'B'$ by reflecting it successively in two lines l_2 and l_1: It is sufficient to choose for l_1 the perpendicular bisector of the segment AA' (if A' coincides with A, then for l_1 one can choose any line passing through A), and for l_2, the bisector of the angle B_1AB where B_1 is the point symmetric to B' with respect to l_1 (Figure 58a). It only remains to use Propositions 2 and 3 of pages 49–50. The proof of Theorem 2 is based on the fact that two equal segments AB and $A'B'$ can be carried into each other by a glide reflection or a reflection. But AB can be carried into $A'B'$ by a sequence of three reflections in lines l_1, l_2, and l_3; the axis l_1 of the first reflection can be chosen completely arbitrarily, and then the lines l_2 and l_3 are chosen so that the sum of the reflections in these two lines carries the segment A_1B_1, obtained from AB by reflection in l_1, into $A'B'$ (Figure 58b). It only remains to use Propositions 4 and 5 of pages 53–54.

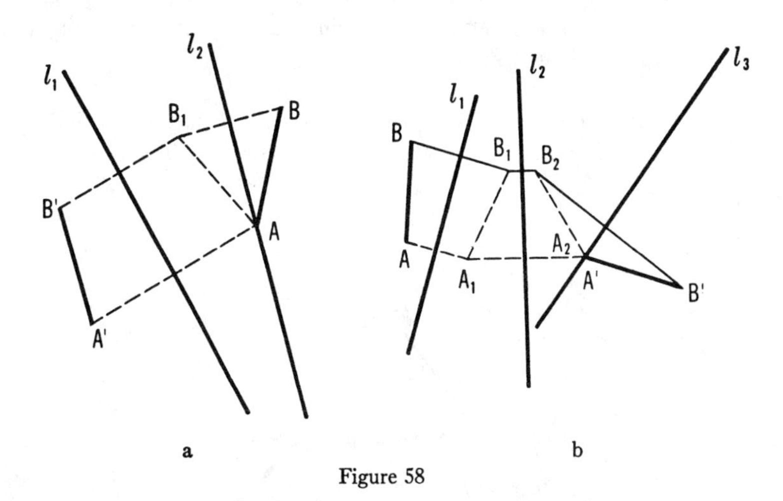

Figure 58

Conversely, all the propositions on the addition of isometries can be derived from Theorems 1 and 2. Thus, Theorem 1 asserts that each pair of directly congruent figures can be obtained from each other by a rotation or a translation. But if two figures F and F' are related by two reflections, or in general by an even number of reflections, then these figures are directly congruent (since a single reflection changes the orientation of a triangle, but two reflections leave it unchanged). Therefore F' may be obtained from F by a rotation or a translation—that is, *the sum of two reflections* (or more generally, of an even number of reflections) *is a rotation or a translation* (see page 55). In a completely analogous manner one shows from Theorem 2 that *the sum of three reflections* (or more generally, of an odd number of reflections) *is a glide reflection or a reflection* (see page 55). From Theorem 1 it also follows that *the sum of two rotations is a rotation or a translation* (see page 34 or the text in small print on pages 51–52), and that *the sum of two glide reflections is a rotation or a translation*, etc.

45. Let two lines l_1 and l_2 be given, together with a point A on the line l_1 and a point B on l_2. Draw a line m, meeting the lines l_1 and l_2 in points X and Y with $AX = BY$, and such that:

(a) The line m is parallel to a given line n.

(b) The line m passes through a given point M.

(c) The segment XY has a given length a.

(d) The segment XY is divided in half by a given line r.

46. Let three lines l_1, l_2, and l_3 be given, together with three points A, B, and C, one on each of the lines. Draw a line m that meets the lines l_1, l_2, and l_3 in points X, Y, and Z such that $AX = BY = CZ$.

47. Let a triangle ABC be given. Draw a line l that meets the sides AB and AC in points P and Q with $BP = PQ = QC$ (Figure 59).

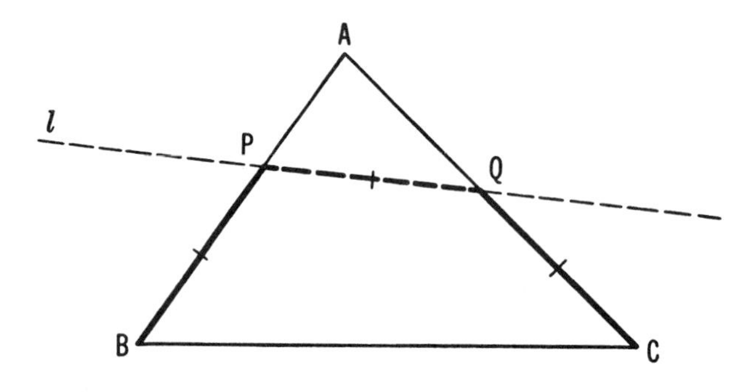

Figure 59

Theorems 1 and 2 can be used as the foundation for a *definition* of isometries in the plane. Indeed, when in geometry we speak of isometries, we are interested only in the result of moving a figure from one position to another, but not in the actual process of movement (the paths described by the individual points of the figures during the movement, the velocities of these points, etc.). And since by Theorems 1 and 2 any two congruent figures can be made to coincide by a translation, or a rotation, or a reflection, or a glide reflection, then in geometry one can say that these four types of isometries exhaust the isometries of the plane.† This listing of all the isometries can serve as the definition of an isometry in the plane. Therefore one can say that geometry studies properties of figures that are not changed by translation, rotation, reflection, and glide reflection (see the introduction, page 10).

In mathematics (and in science in general) one encounters two different types of definitions. A new concept may be defined by listing the properties it is to have; thus, for example, parallel lines in the plane can be defined as lines that never meet, no matter how far they are extended; an arithmetic progression may be defined as a sequence of numbers with the property that the difference of any two consecutive numbers of the sequence has a fixed value; a steam engine may be defined to be a mechanism transforming heat energy into mechanical energy. Definitions of this type are called *descriptive*. One may also define a new object by directly indicating how to construct it, rather than by an enumeration of its properties. Thus, parallel lines may be defined as two perpendiculars to one and the same line (here one gives a method for constructing parallel lines); an arithmetic progression is the sequence of numbers

$$a, \quad a + d, \quad a + 2d, \quad a + 3d, \quad \cdots$$

(where the number a is called the first term, and d is called the difference of the progression); a definition of a steam engine may consist in giving a description of its construction. Definitions of this sort are called *constructive*. One could say that the fundamental task of science is to find constructive definitions for concepts that have previously had only descriptive definitions. Thus the problem of creating a steam engine can be considered as the problem of starting with the descriptive definition, as a mechanism converting heat energy into mechanical energy, and finding a constructive definition, that is, actually building it.‡

† In contrast to this, in mechanics, where one studies the process of movement, it is impossible to give such a simple enumeration of all motions in the plane.

‡ Let us note also that finding a constructive definition for an object that previously had only a descriptive definition serves at the same time as a *proof of the existence* of this object; the existence of the object does not follow at all from a descriptive definition alone. Examples of descriptive definitions that do not correspond to any real object are the following: "A tricornicum is a triangle in which two angle bisectors are perpendicular" [compare this with the solution to Problem 26(a) of this chapter], or "a perpetual motion machine is a mechanism that is able to accomplish work without the use of energy"; a constructive definition is clearly impossible in these cases.

The definition of isometry as a transformation that does not change the distances between points (see the introduction to this volume, page 11) is a typical descriptive definition. And the basic problem in the theory of isometries is to find a constructive definition of isometry, that is, to enumerate all the isometries of the plane. And it is precisely this problem that is solved by Theorems 1 and 2 of this section; these are therefore the basic results of this section.

Conversely, given a constructive definition of some concept, it is often convenient to find a simple descriptive definition which may be useful in studying the properties of the new object. We have also had examples of this sort in the present chapter. Thus, after the first constructive definition of translation, we also gave a purely descriptive definition of this transformation: A translation is a transformation of the plane in which each segment AB is taken into a segment $A'B'$ that is equal to, parallel to, and has the same direction as the segment AB (see pages 18–19). This definition is very useful in solving the problem of what sort of transformation is given by the sum of two translations; the first (constructive) definition of translation would have been less useful in solving this problem. In the same way the solution of the problem of finding the transformation resulting from the sum of two rotations is based on the descriptive definition of rotation: A rotation is a transformation of the plane in which each segment AB is carried into a segment $A'B'$ equal to AB and forming a known angle α with AB (see page 32). The reader himself should seek other such examples in this book.

Solutions

Chapter One. Displacements

1. Translate the circle S_1 a distance a in the direction l, and let S_1' be its new position; let A' and B' be the points of intersection of S_1' with the circle S_2 (see Figure 60). The two lines parallel to l, one through the point A' and the other through the point B' will each solve the problem (the segments AA' and BB' in Figure 60 are each equal to the distance a of the translation). One can find two additional solutions by translating S_1 in the opposite direction a distance a parallel to l into the new position S_1''.

Depending on the number of points of intersection of the circles S_1' and S_1'' with S_2, the problem may have infinitely many solutions, four solutions, three solutions, two solutions, one solution, or no solution at all. In the case shown in Figure 60 the problem has three solutions.

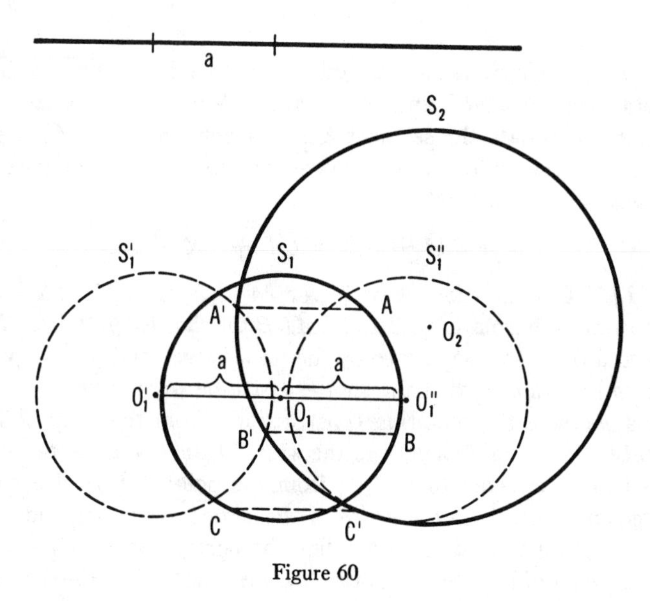

Figure 60

2. (a) Assume that the problem has been solved, and translate the segment MN into a new position AN' in such a manner that the point M is carried into the point A (Figure 61a). Then $AM = N'N$, and therefore

$$AM + NB = N'N + NB.$$

Thus the path $AMNB$ will be the shortest path if and only if the points N', N, and B lie on one line.

Thus we have the following construction: From the point A lay off a segment AN' equal in length to the width of the river, perpendicular to the river, and directed toward it; pass a line through the points N' and B; let N be the point of intersection of this line with the river bank nearest to B; build the bridge across the river at the point N.

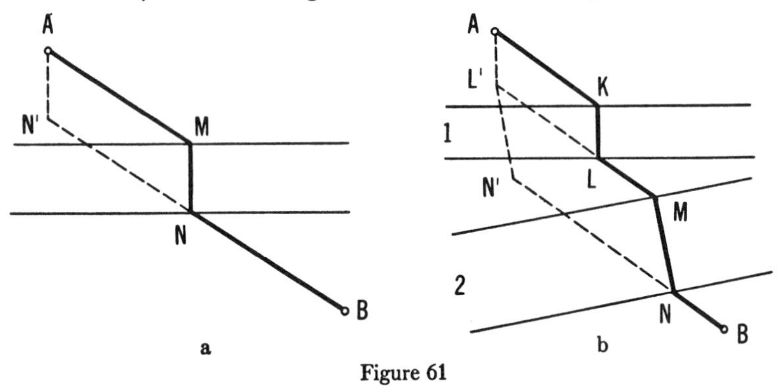

Figure 61

(b) For simplicity we consider the case of two rivers. Assume that the problem has been solved, and let KL and MN be the two bridges across the rivers. Translate the segment KL to a new position AL' in such a manner that the endpoint K is taken into the point A (Figure 61b). Then $AK = L'L$ and

$$AK + LM + NB = L'L + LM + NB.$$

If $AKLMNB$ is the shortest path from A to B, then $L'LMNB$ will be the shortest path from L' to B and $LMNB$ the shortest path from L to B. But L and B are only separated by the second river, and so from part (a) we know how to construct the shortest path between them.

Thus we have the following construction: From the point A lay off a segment AL' equal in length to the width of the first river, perpendicular to it, and directed toward it; from the point L lay off a segment $L'N'$ equal in length to the width of the second river, perpendicular to it, and directed toward it. Pass a line through the points N' and B; let N be the point of intersection of this line with the bank of the second river nearest to B. The bridge across the second river should be built at N. Let M be the other endpoint of this bridge. Pass a line through the point M parallel to the line $N'B$, and let L be the point of intersection of this line with the bank of the first river nearest to M. The first bridge should be built at L.

3. (a) Let M be a point in the plane for which $MP + MQ = a$, where P and Q are the feet of the perpendiculars from M to the lines l_1 and l_2, respectively (Figure 62a). Translate the line l_2 a distance a in the direction QM. If l_2' is the new line obtained by this translation, then it is clear that the distance MQ' of the point M from the line l_2' is equal to $a - MQ = MP$. Consequently M is on the bisector of one of the angles between the lines l_1 and l_2'.

From this it is clear that all points of the desired locus lie on the bisectors of the angles formed by the line l_1 with the lines l_2' and l_2'', obtained from l_2 by translation through a distance a in the direction perpendicular to l_2. However, not all the points on these four bisectors are points of our locus. From Figure 62a it is not difficult to see that only the points on the rectangle $ABCD$ formed by the intersections of the four bisectors will be points of the locus.

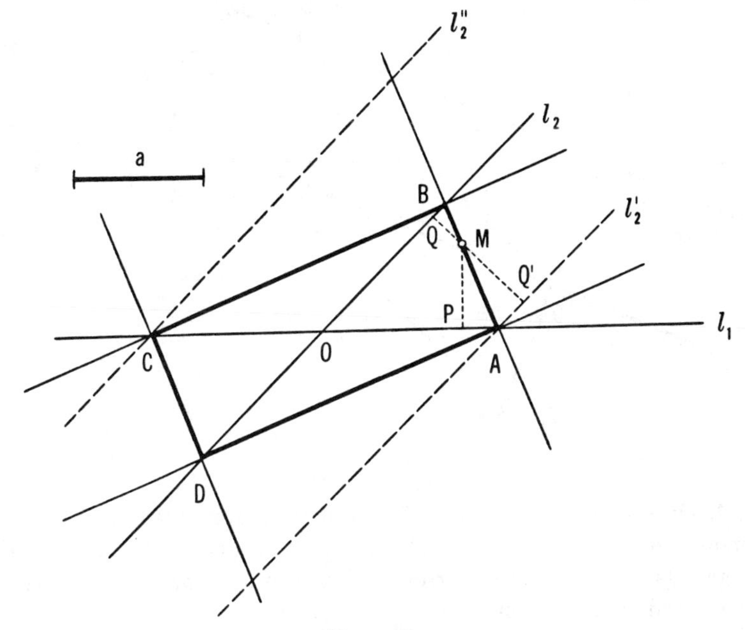

Figure 62a

(b) Let M be a point of the plane satisfying one of the following two equations:

$$MP - MQ = a \quad \text{or} \quad MQ - MP = a,$$

where P and Q are the feet of the perpendiculars from M to the lines l_1 and l_2 (in Figure 62b, the point M satisfies the second equation). Translate the line l_2 a distance a in the direction QM, and let l_2' be the new line. Just as in part (a) one can show that M is equidistant from l_1 and l_2' (see Figure 62b, where $MQ - MP = a$, $M_1P_1 - M_1Q_1 = a$). It follows that all points of the desired locus lie on the bisectors of the four angles formed by the line l_1 with the lines l_2' and l_2''; however in the present case only points lying on the *extensions* of the sides of the rectangle $ABCD$ will be points of the locus (the equation $MP - MQ = a$ is satisfied by the points on HBG and LDN, while the equation $MQ - MP = a$ is satisfied by the points on EAF and ICK).

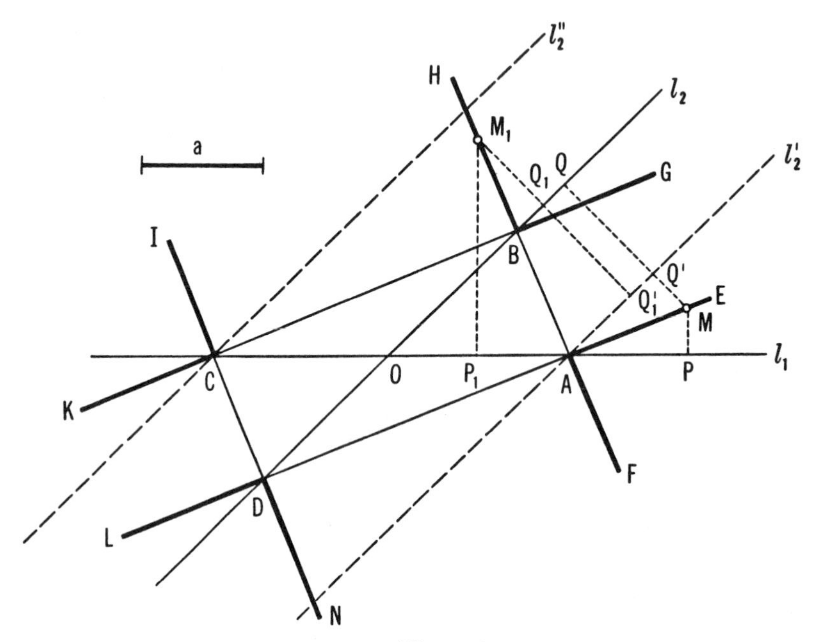

Figure 62b

4. Observe that triangle BDE is obtained from triangle DAF by a translation (in the direction AB through a distance AD); thus the line segments joining pairs of corresponding points in these two figures are equal and parallel to one another. Therefore

$$O_1O_2 = Q_1Q_2, \qquad O_1O_2 \parallel Q_1Q_2.$$

Similarly one has

$$O_2O_3 = Q_2Q_3, \qquad O_2O_3 \parallel Q_2Q_3,$$

and

$$O_3O_1 = Q_3Q_1, \qquad O_3O_1 \parallel Q_3Q_1.$$

Therefore triangles $O_1O_2O_3$ and $Q_1Q_2Q_3$ are congruent (in fact, their corresponding sides are parallel, that is, the triangles are obtained from one another by a translation—see pages 18–19).

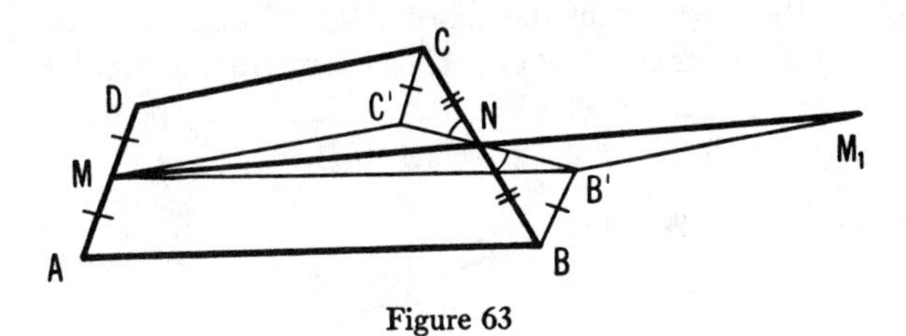

Figure 63

5. Translate the sides AB and DC of the quadrilateral $ABCD$ into the new positions MB' and MC' (Figure 63). The two quadrilaterals $AMB'B$ and $DMC'C$ thus formed will be parallelograms, and therefore

$$BB' \parallel AM \quad \text{and} \quad BB' = AM,$$
$$CC' \parallel DM \quad \text{and} \quad CC' = DM.$$

But $AM = MD$ (M is the midpoint of side AD); thus the segments BB' and CC' are equal and parallel. Since, in addition, $BN = NC$, it follows that

$$\triangle BNB' \cong \triangle CNC'.$$

Therefore $B'N = NC'$ and $\measuredangle BNB' = \measuredangle CNC'$, that is, the segments $B'N$ and NC' are extensions of each other.

Thus we have constructed a triangle $MB'C'$ in which, by the conditions of the problem, the median MN is equal to half the sum of the two adjacent sides MB' and MC' (since $MB' = AB$, $MC' = DC$). If we extend the median MN past the point N a distance $NM_1 = MN$ and join M_1 with B', we obtain a triangle MM_1B' in which the side $MM_1 = 2MN$ is equal to the sum of the sides MB' and $B'M_1 = MC'$, which is impossible. Consequently the point B must lie on the segment MM_1. But this means that

$$MB' \parallel MN \parallel MC';$$

therefore

$$AB \parallel MN \quad \text{and} \quad DC \parallel MN,$$

that is, the quadrilateral $ABCD$ is a trapezoid.

6. Assume that the problem has been solved. Translate the segment AX a distance $EF = a$ in the direction of the line CD, and let the new position be $A'X'$ (Figure 64).

Clearly $A'X'$ passes through the point F. Further

$$\measuredangle A'FB = \measuredangle AXB = \tfrac{1}{2}AmB;\dagger$$

therefore we may regard the angle $A'FB$ as known.

Thus we have the following construction: Translate the point A a distance a in the direction of the chord CD, and denote its new position by A'. Using the segment $A'B$ as a chord, construct a circular arc[T] that subtends an angle equal to $\measuredangle AXB$ (that is, if Y is any point on the circular arc, then $\measuredangle A'YB = \measuredangle AXB = \tfrac{1}{2}AmB$).

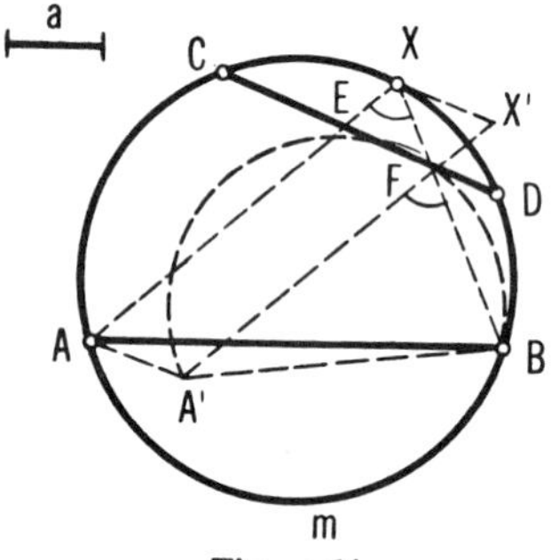

Figure 64

If this circular arc intersects the chord CD in two points, either one of them may be taken as the point F, and the point X is obtained as the point of intersection of the original circle with the line BF. In this case the problem has two solutions.

If the circular arc is tangent to CD, the point of tangency must be taken as the point F, and the problem has just one solution.

If the arc does not intersect CD at all, the problem has no solution.

If one assumes that CD is intersected by the extensions of chords AX and BX (and that points E and F are outside the circle—on the extension of chord CD), then the problem can have up to four solutions. (This is due to the fact that A may be translated in either of two opposite directions.)[TT]

† AmB stands for arc AmB.

[T] For the details of this construction, see, for example, *Hungarian Problem Book 1* in this series, Problem 1895/2, Note.

[TT] The foregoing paragraphs concerning the number of solutions were added in translation.

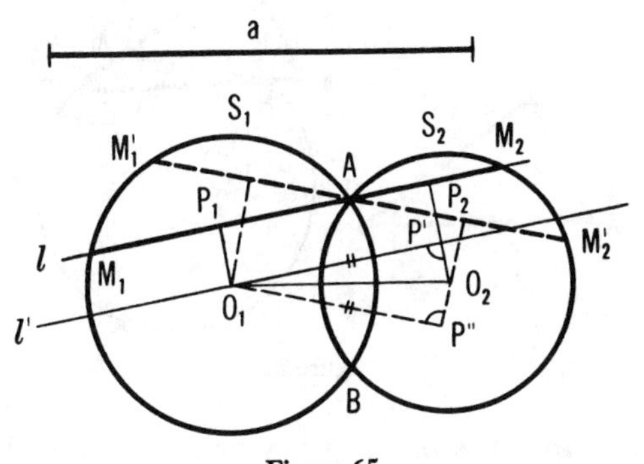

Figure 65

7. (a) Assume that the problem has been solved, i.e., that $M_1M_2 = a$ (Figure 65). From the centers O_1 and O_2 of the circles S_1 and S_2, drop perpendiculars O_1P_1 and O_2P_2 onto the line l; then

$$AP_1 = \tfrac{1}{2}AM_1, \qquad AP_2 = \tfrac{1}{2}AM_2,$$

and consequently,

$$P_1P_2 = \tfrac{1}{2}(AM_1 + AM_2) = \tfrac{1}{2}M_1M_2 = \tfrac{1}{2}a.$$

Translate the line l into a line l' passing through the point O_1; let P' be the point of intersection of l' with the line O_2P_2. Then

$$O_1P' = P_1P_2 = \tfrac{1}{2}a,$$

since the quadrilateral $P_1O_1P'P_2$ is a rectangle.

Thus we have the following construction: Construct a right triangle O_1O_2P' with O_1O_2 as hypotenuse and with side $O_1P' = \frac{1}{2}a$. The desired line l will be parallel to the line O_1P'.

If $O_1O_2 > \frac{1}{2}a$ the problem has two solutions (the construction of a second solution to the problem is indicated in dotted lines in Figure 65); if $O_1O_2 = \frac{1}{2}a$ there is one solution, and if $O_1O_2 < \frac{1}{2}a$ there are no solutions.

(b) Let M, N, P be the three given points and let ABC be the given triangle (Figure 66). On the segments MN and MP construct circular arcs subtending angles equal to $\measuredangle ACB$ and $\measuredangle ABC$, respectively. Thus we are led to the following problem: Pass a line B_1C_1 through the point M in such a way that the segment cut off by the two circular arcs has length BC, that is, we are led to Problem (a). The problem may have two solutions, or one solution, or no solutions at all (depending on which sides of the triangle are to pass through each of the three given points).

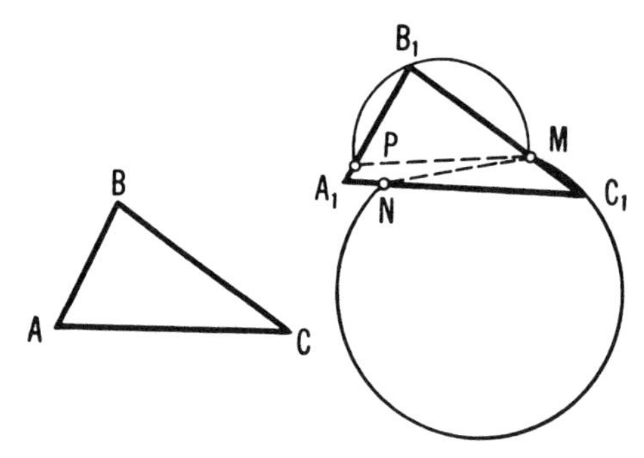

Figure 66

8. (a) Assume that the problem has been solved, and let the line l meet the circles S_1 and S_2 in points A, B and C, D (Figure 67a). Translate the circle S_1 a distance AC in the direction of the line l, and let S_1' be its new position. Since $AB = CD$, the segment AB will coincide with CD; therefore the centers O_2 and O_1' of the circles S_2 and S_1' will both lie on the perpendicular bisector of the segment CD.

Thus we have the following construction: Let m be the line perpendicular to l_1 and passing through the center O_2 of the circle S_2; let n be the line parallel to l_1 and passing through the center O_1 of the circle S_1; Let O_1' be the point of intersection of these two lines. Translate S_1 into a new position S_1' with center at O_1'. The line through the points of intersection of S_2 and S_1' is the solution to the problem.

The problem can have one solution or no solution.

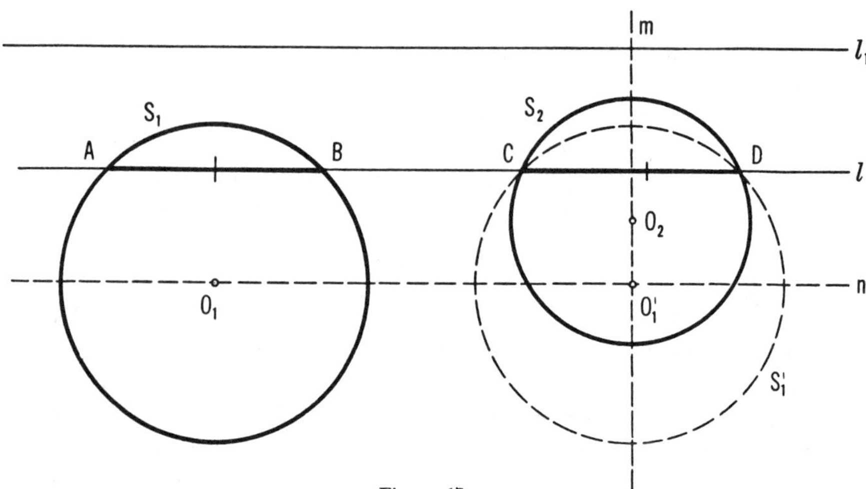

Figure 67a

(b) Assume that the problem has been solved and let the line l meet S_1 and S_2 in points A, B and C, D; then $AB + CD = a$ (Figure 67b). Translate the circle S_1 a distance a in the direction of l and denote its new position by S_1'; then

$$AA' = a = AB + CD,$$

that is, $BA' = CD$. Therefore, if we translate the circle S_2 in the direction of l into a new position S_2' whose center O_2' is on the perpendicular bisector m of the segment O_1O_1' (O_1 and O_1' are the centers of the circles S_1 and S_1'), then the chord CD of S_2 will be taken into BA'.

Thus we have the following construction: Translate the circle S_1 a distance a in the direction of the line l_1, and denote the new position by S_1'; then translate S_2 in the direction of l_1 into a new position S_2' whose center lies on the perpendicular bisector m of the segment O_1O_1'. The points of intersection of the circles S_1 and S_2' (in the diagram they are the points B and B_1) determine the desired lines. The problem has at most two solutions; the number of solutions depends upon the number of points of intersection of the circles S_1 and S_2' (a case when there are two solutions l and l' is shown in Figure 67b).

The other part of the problem, where the *difference* of the two chords cut off on the line l by the two circles is given, can be solved in a similar manner.

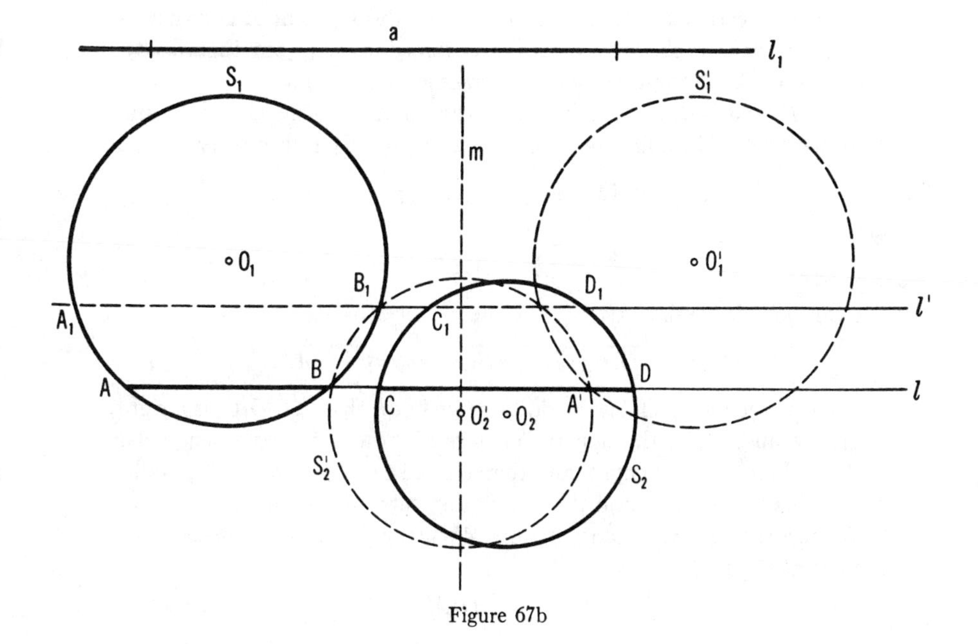

Figure 67b

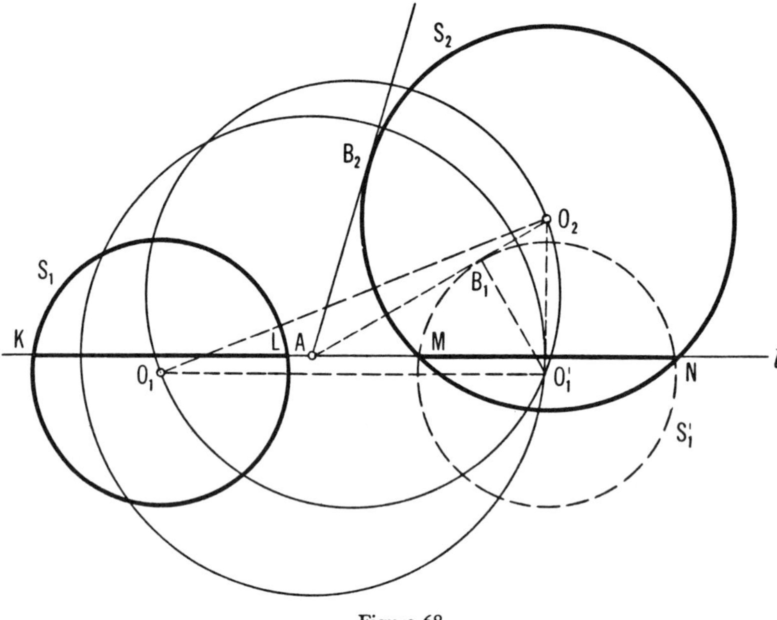

Figure 68

(c) Assume that the problem has been solved, and translate the circle S_1 in the direction of the line KN so that the segment KL coincides with MN; denote the new circle so obtained by S_1' (see Figure 68). Thus the circles S_2 and S_1' have the common chord MN.

Let AB_1 and AB_2 be tangents from the point A to the circles S_1' and S_2 respectively (the points of tangency are B_1 and B_2, respectively). Then

$$(AB_1)^2 = AM \cdot AN; \qquad (AB_2)^2 = AM \cdot AN$$

and therefore

$$(AB_1)^2 = (AB_2)^2.$$

We can now determine AO_1' (O_1' is the center of S_1'):

$$AO_1' = \sqrt{(O_1'B_1)^2 + (AB_1)^2} = \sqrt{r_1^2 + (AB_2)^2},$$

where r_1 is the radius of S_1; in addition, we know that $\measuredangle O_1O_1'O_2$ is a right angle, because $O_1'O_2$, through the centers of S_1' and S_2, is perpendicular to MN, their common chord, and therefore also to O_1O_1', which is parallel to l. This enables us to find the translation carrying S_1 into S_1'.

We use the following construction. With the point A as center, draw a circle of radius

$$\sqrt{r_1^2 + (AB_2)^2};$$

draw a second circle having the segment O_1O_2 as diameter. The intersection of these two circles determines the position of the center O_1' of the circle S_1' of radius r_1. Now find the points M and N of intersection of the circles S_2 and S_1' and draw the line MN, which will be the solution to the problem. Indeed, the point A lies on the line MN; for otherwise the equation $(AB_1)^2 = (AB_2)^2$ could not be satisfied [if the line AM were to intersect the circles S_2 and S_1' in distinct points N_2 and N_1, then we would have $(AB_2)^2 = AM \cdot AN_2$ and $(AB_1)^2 = AM \cdot AN_1$]. Also, O_2O_1' is perpendicular to MN, and O_1O_1' is perpendicular to O_2O_1'; therefore $O_1O_1' \parallel MN$, that is, the chords KL and MN of the circles S_1 and S_1' are at the same distance from the centers O_1 and O_1'. But this means that the chords KL and MN have the same length, which was to be proved.

The problem has at most two solutions.

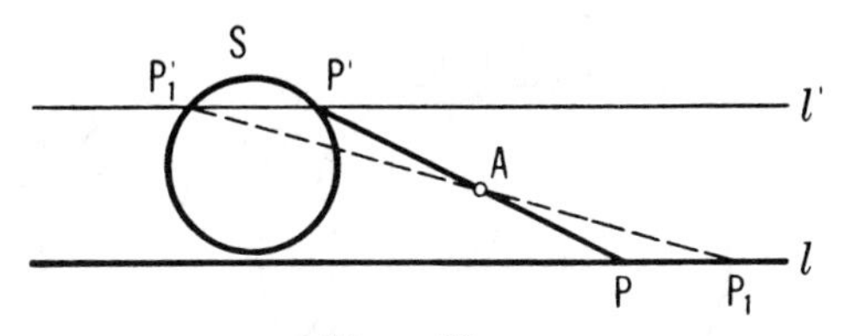

Figure 69

9. Draw the line l' obtained from l by a half turn about the point A (Figure 69); let P' be one of the points of intersection of this line with the circle S. Then the line $P'A$ is a solution to the problem, since the point P of intersection of this line with the line l is obtained from P' by a half turn about A, and therefore $P'A = AP$.

There are at most two solutions to this problem.

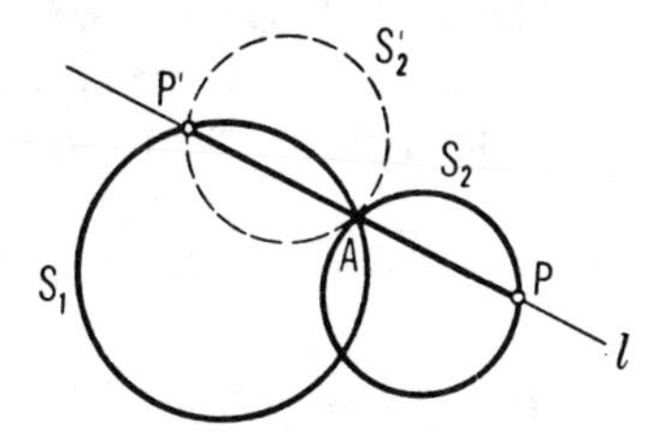

Figure 70a

10. (a) Draw the circle S_2' obtained from S_2 by a half turn about the point A (Figure 70a). The circles S_1 and S_2' intersect in the point A; let P' be their other point of intersection. Then the line $P'A$ will solve the problem, because the point P where this line meets the circle S_2 is obtained from P' by a half turn about A, and therefore $P'A = AP$.

If the circles S_1 and S_2 intersect in two points, then the problem has exactly one solution; if they are tangent, then there is no solution if the radii are different, and there are infinitely many solutions if the radii are equal.

Remark: This problem is a special case of Problem 8(c), and it has a much simpler solution.

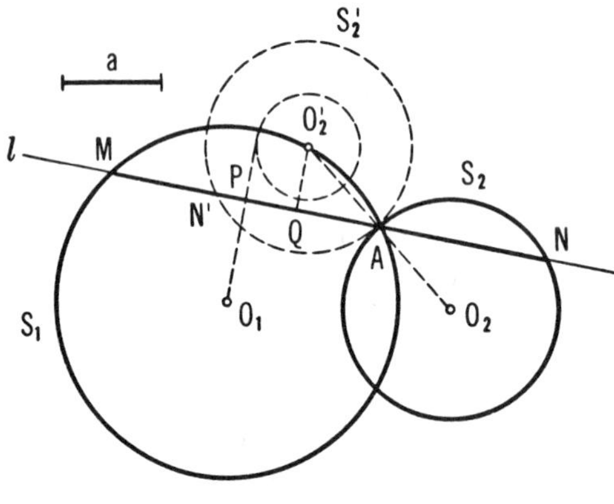

Figure 70b

(b) Draw the circle S_2' obtained from S_2 by a half turn about the point A. Assume that the problem has been solved and that the line MAN is the solution (Figure 70b). Let N' be the point where this line intersects the circle S_2'; then $MN' = a$. From the centers O_1 and O_2' of the circles S_1 and S_2', drop perpendiculars O_1P and $O_2'Q$ to the line MAN; then

$$PA = \tfrac{1}{2}MA, \qquad QA = \tfrac{1}{2}N'A$$

and

$$PQ = PA - QA = \tfrac{1}{2}(MA - N'A) = \tfrac{1}{2}a.$$

Thus the distance from the point O_2' to the line O_1P is equal to $\frac{1}{2}a$, that is, the line O_1P is tangent to the circle with center O_2' and radius $\frac{1}{2}a$. This enables us now to find the line O_1P without assuming that the solution to the whole problem is already known. Having found O_1P we can now easily construct $MAN \perp O_1P$.

There are at most two solutions to the problem.

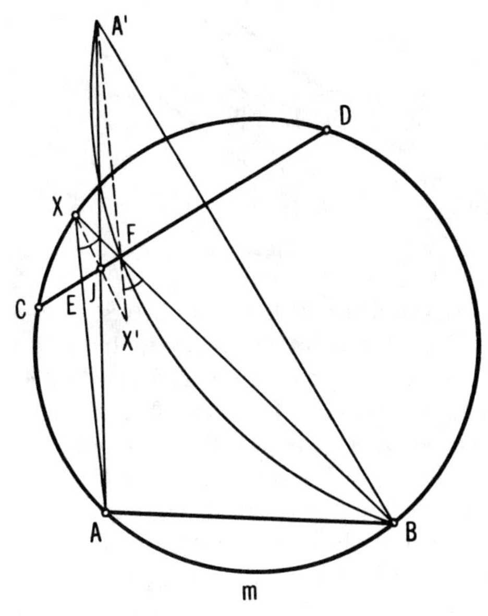

Figure 71

11. Assume that the problem has been solved (Figure 71), and let $A'X'$ be the segment obtained from AX by a half turn about the point J. Since AX passes through E, $A'X'$ will pass through F. Since $X'A' \parallel AX$, we see that

$$\sphericalangle X'FB = \sphericalangle AXB = \tfrac{1}{2}AmB;$$

therefore, $\sphericalangle A'FB = 180° - \sphericalangle X'FB$ and so we may regard

$$\sphericalangle A'FB = 180° - \tfrac{1}{2}AmB$$

as known.

Thus we have the following construction: Let A' be the point obtained from A by a half turn about J. On the segment $A'B$ construct the circle arc that subtends an angle of

$$180° - \tfrac{1}{2}AmB.$$

The point of intersection of this arc with the chord CD determines the point F, and the other intersection of the line BF with the circumference is the desired point X.

The problem has a unique solution; if one assumes that CD is intersected by the extensions of chords AX and BX, then there may be two solutions (cf. solution of Problem 6).

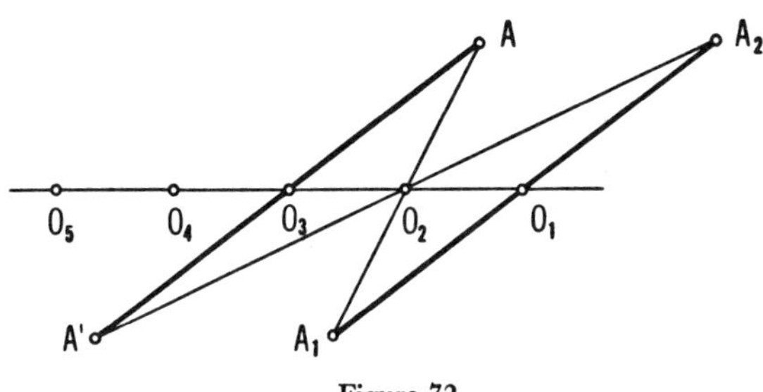

Figure 72

12. Assume that the figure F has two centers of symmetry, O_1 and O_2 (Figure 72). Then the point O_3, obtained from O_1 by a half turn about O_2 is also a center of symmetry of F. Indeed, if A is any point of F, then the points A_1, A_2, and A', where A_1 is obtained from A by a half turn about O_2, A_2 from A_1 by a half turn about O_1, and A' from A_2 by a half turn about O_2, will also be points of F (since O_1 and O_2 are centers of symmetry). But the point A' is also obtained from A by a half turn about O_3; indeed, the segments AO_3 and O_3A' are equal, parallel, and have opposite directions, since the pairs of segments AO_3 and A_1O_1, A_1O_1 and A_2O_1, A_2O_1 and $A'O_3$ are equal, parallel, and have opposite directions.

Thus if A is any point of F, then the symmetric point A' obtained from A by a half turn about O_3 is also a point of F, that is, O_3 is a center of symmetry of F.

Similarly one shows that the point O_4, obtained from O_2 by a half turn about O_3, and the point O_5, obtained from O_3 by a half turn about O_4, etc. are centers of symmetry. Thus we see that if the figure F has two distinct centers of symmetry then it has infinitely many.

13. (a) The segment A_nB_n is obtained from AB by n successive half turns about the points $O_1, O_2, \cdots, O_n$ (n even). But the sum of the half turns about O_1 and O_2 is a translation; the sum of the half turns about O_3 and O_4 is a translation; the sum of the half turns about O_5 and O_6 is a translation; $\cdots$; finally, the sum of the half turns about O_{n-1} and O_n is also a translation. Therefore A_nB_n is obtained from AB by $\frac{1}{2}n$ successive translations. Since any sum of translations is again a translation the segment A_nB_n is obtained from AB by a translation, and therefore $AA_n = BB_n$.

If n is odd the assertion of the problem is false, because the sum of an odd number of half turns is a translation plus a half turn, or, what is the same thing, is a half turn about some other point (see page 34); therefore, in general $AA_n \neq BB_n$ (although $AB_n = BA_n$).

(b) Since the sum of an odd number of half turns is a half turn [see the solution to Problem (a)], the point A_n obtained from A by the n successive half turns about the points $O_1, O_2, \cdots, O_n$ can also be obtained from A by a single half turn about some point O. The point A_{2n} is obtained from A_n by these same n half turns; therefore it can also be obtained from A_n by the single half turn about the point O. But this means that A_{2n} coincides with A.

If n is even then A_n is obtained from A by a translation, and A_{2n} is obtained from A_n by this same translation; therefore A_{2n} will not, in general, coincide with A. (It will coincide with A if this translation is the identity transformation, i.e., a translation through zero distance.[T])

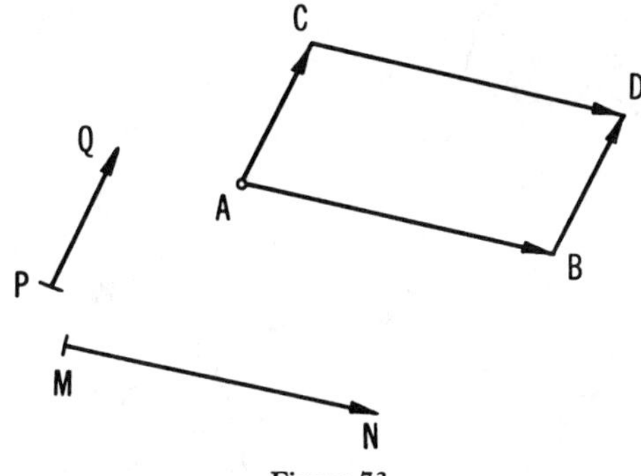

Figure 73

14. (a) The sum of the two half turns about the points O_1 and O_2 is a translation (see page 25) and the sum of the half turns about the points O_3 and O_4 is another translation (in general, different from the first). Thus the "first" point A_4 is obtained from A by performing two translations in succession; the "second" point (we denote it by A_4') is obtained from A by performing the same two translations *in the opposite order*. But *the sum of two translations is independent of the order in which they are performed.* (To prove this it is sufficient to consider Figure 73, where the points B and C are obtained from the point A by the translations indicated by the segments MN and PQ respectively. The point D is obtained from the point B by the translation PQ, and D is also obtained from C by the translation MN. From this the assertion of the theorem follows.)

(b) This problem is clearly the same as Problem 13(b) (for $n = 5$), since Problem 13(b) tells us that the point A_5, obtained from A by five successive half turns about the points O_1, O_2, O_3, O_4, O_5, is taken back into the point A by these same five half turns performed in the same order.

[T] This sentence was added in translation.

(c) Whenever n is odd, the final positions will be the same (see Problem 13).

[The two points obtained by the n half turns will also coincide in case $n = 2k$ is an even number and there exists a k-gon $M_1M_2\cdots M_k$, whose sides M_1M_2, M_2M_3, $\cdots$, M_kM_1 are equal to, parallel to, and have the same direction as the segments O_1O_2, O_3O_4, $\cdots$, $O_{n-1}O_n$ (in this case the sum of the n half turns about the points O_1, O_2, $\cdots$, O_n, carried out in either this order or in the reverse order, is a "translation through zero distance", that is, it is the identity transformation).]

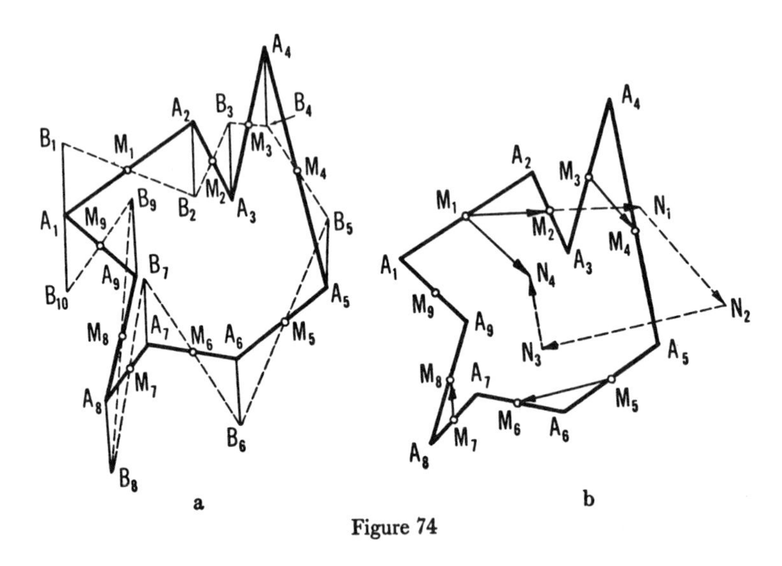

Figure 74

15. *First solution.* Assume that the problem has been solved and let

$$A_1A_2\cdots A_9$$

be the nine-gon, with M_1, M_2, $\cdots$, M_9 the centers of its sides (Figure 74a; here we are taking $n = 9$). Let B_1 be any point in the plane and let B_2 be obtained from it by a half turn about M_1. Let B_3 be obtained from B_2 by a half turn about M_2. Continue this until finally B_{10} is obtained from B_9 by a half turn about M_9. Since each of the segments A_2B_2, A_3B_3, $\cdots$, A_1B_{10} is obtained from the preceding one by a half turn, they are all parallel and have the same length, and each one has a direction opposite to the direction of the one before it. Therefore A_1B_1 and A_1B_{10} are equal and parallel and have opposite directions, which means that the point A_1 is the midpoint of the segment B_1B_{10}. This enables us to find A_1, since by starting with any point B_1 we can find B_{10}. The remaining vertices A_2, A_3, $\cdots$, A_9 are then found by successive half turns about M_1, M_2, $\cdots$, M_9.

The problem always has a unique solution; however, the nine-gon that is obtained need not be convex and may even intersect itself.

If n is even and if we repeat the same reasoning as before, i.e., if we assume that the problem has been solved, we see that A_1B_{n+1} and A_1B_1 are equal, parallel and have the same direction, that is, they coincide. Therefore if B_{n+1} does not coincide with B_1, then the problem has no solution. If B_{n+1} does coincide with B_1 then A_1B_1 will coincide with A_1B_{n+1} no matter where the point A_1 is chosen. In this case there are infinitely many solutions; any point in the plane can be taken for the vertex A_1.

Second solution. The vertex A_1 of the desired n-gon will be taken into itself by the sum of the half turns about the points $M_1, M_2, \cdots, M_n$, that is, A_1 is a fixed point of the sum of these n half turns (see Figure 74b) where the case $n = 9$ is shown). If n were even then the sum of n half turns would be a translation [see the solution to Problem 13(a)]. Since a translation has no fixed points, it follows that for n even the problem has, in general, no solution. The only exception occurs when the sum of the n half turns is the identity transformation (a translation through zero distance), which leaves all points in the plane fixed; in this case the problem has infinitely many solutions; any point in the plane can be taken for the vertex A_1.† If n is odd (for example, $n = 9$), then the sum of n half turns is a half turn. Since a half turn has exactly one fixed point, namely the center of symmetry, it follows that the vertex A_1 of the desired nine-gon must coincide with this center of symmetry; in this case the problem has a unique solution.

We now show how to construct the center of symmetry of the sum of the nine half turns about the points $M_1, M_2, \cdots, M_9$. The sum of the half turns about M_1 and M_2 is a translation in the direction M_1M_2 through a distance $2M_1M_2$; the sum of the half turns about M_3 and M_4 is a translation in the direction M_3M_4 through a distance $2M_3M_4$, etc. Thus the sum of the first eight half turns is the same as the sum of the four translations in the directions M_1M_2 (or M_1N_1), M_3M_4 ($\parallel N_1N_2$), M_5M_6 ($\parallel N_2N_3$) and M_7M_8 ($\parallel N_3N_4$) through distances $2M_1M_2$ ($= M_1N_1$), $2M_3M_4$ ($= N_1N_2$), $2M_5M_6$ ($= N_2N_3$), and $2M_7M_8$ ($= N_3N_4$) respectively (see Figure 74b), which is a single translation in the direction M_1N_4 through a distance M_1N_4. The point A_1 is the center of symmetry of the half turn that is the sum of a translation in the direction M_1N_4 through a distance M_1N_4 and a half turn about the point M_9. To find A_1 it is sufficient to lay off a segment M_9A_1 starting at M_9, parallel to N_4M_1 and of length $\frac{1}{2}M_1N_4$ (Figure 74b; compare this with Figure 18). Having found A_1, we have no difficulty in finding the remaining vertices of the nine-gon.

† See the note at the end of the solution of Problem 16(b) for a discussion of the conditions that the points $M_1, M_2, \cdots, M_n$ must satisfy in this case.

16. (a) If M, N, P, and Q are the midpoints of the sides of the quadrilateral $ABCD$ (see Figure 22a), then four half turns performed in succession about the points M, N, P, and Q will carry the point A into itself (compare with the solution to Problem 15). Now this is possible only in case the sum of the four half turns about the points M, N, P, and Q, which is equal to the sum of two translations in the directions MN and PQ through distances $2MN$ and $2PQ$ respectively, is the identity transformation. But this means that the segments MN and PQ are parallel, equal in length and oppositely directed, that is, the quadrilateral $MNPQ$ is a parallelogram.

(b) Just as in part (a), we conclude that the sum of the translations in the directions M_1M_2, M_3M_4, and M_5M_6 through distances $2M_1M_2$, $2M_3M_4$, and $2M_5M_6$ is the identity transformation. Therefore there is a triangle whose sides are parallel to M_1M_2, M_3M_4, and M_5M_6, and equal to $2M_1M_2$, $2M_3M_4$, and $2M_5M_6$; but this means that there is also a triangle whose sides are parallel to and have the same lengths as the segments M_1M_2, M_3M_4, M_5M_6.

In the same way one proves that there exists a triangle whose sides are parallel to, and have the same lengths as the segments M_2M_3, M_4M_5, M_6M_1.

Remark: Using the same method that was used in the solution of Problem 16(b) one can show that a set of $2n$ points $M_1, M_2, \cdots, M_{2n}$ will be the midpoints of the sides of some $2n$-gon if and only if there exists an n-gon whose sides are parallel to and have the same lengths as the segments M_1M_2, $M_3M_4, \cdots, M_{2n-1}M_{2n}$; there will then also exist an n-gon whose sides are parallel to and have the same lengths as M_2M_3, $M_4M_5, \cdots, M_{2n}M_1$.

17. Rotate the line l_1 about the point A through an angle α, and let l_1' denote the new position of the line. Let M be the point of intersection of l_1' with the line l_2 (Figure 75). The circle having its center at A and passing through the point M will solve the problem, since the point of intersection M' of this circle with the line l_1 is taken into the point M by our rotation (that is, the central angle $MAM' = \alpha$).

The problem has two solutions (corresponding to rotations in the two directions), provided that neither of the angles between the lines l_1 and l_2 is equal to α; it has either exactly one solution or infinitely many solutions if one of the angles between the lines l_1 and l_2 is equal to α; it has either no solutions at all or infinitely many solutions if l_1 and l_2 are perpendicular and $\alpha = 90°$.

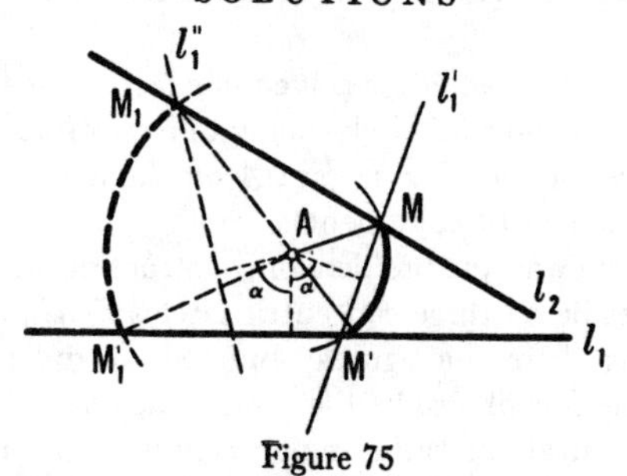

Figure 75

18. Assume that the problem has been solved and let ABC be the desired triangle whose vertices lie on the given lines l_1, l_2, and l_3 (Figure 76). Rotate the line l_2 about the point A through an angle of 60° in the direction from B to C; this will carry the point B into the point C.

Thus we have the following construction: Choose an arbitrary point A on the line l_1 and rotate l_2 about A through an angle of 60°. The point of intersection of the new line l_2' with l_3 is the vertex C of the desired triangle. The problem has two solutions since l_2 can be rotated through 60° in either of two directions; however, these two solutions are congruent.

The problem of constructing an equilateral triangle whose vertices lie on three given concentric circles is solved analogously.

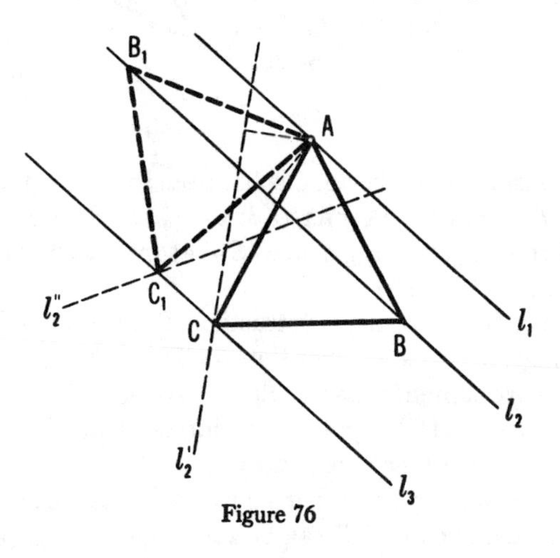

Figure 76

Remark: If we had chosen a different point A' instead of A on the line l_1, then the new figure would be obtained from Figure 76 by an isometry (more precisely, by a translation in the direction l_1 through a distance AA'). But in geometry we do not distinguish between such figures (see the introduction). For this reason we do not consider that the solution to the problem depends on the position of the point A on l_1. If the three lines l_1, l_2, and l_3 were not parallel, then the problem would be solved in exactly the same way; however now we would have to allow infinitely many different solutions corresponding to the different ways of choosing a point A on the line l_1 (since the triangles obtained would no longer be congruent).

In exactly the same way the problem of constructing an equilateral triangle ABC whose vertices lie on three concentric circles S_1, S_2, and S_3 can have at most four solutions (here the figures obtained by different choices of the point A on the circle S_1 will also be the same—they are all obtained from one another by a rotation about the common center of the three circles S_1, S_2, and S_3). On the other hand, if the circles S_1, S_2, and S_3 are not concentric, then the problem will have infinitely many solutions (different choices of the point A on the circle S_1 will correspond to essentially different solutions).

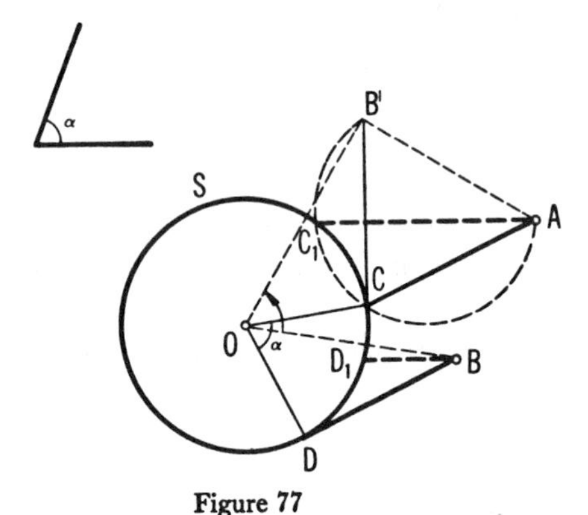

Figure 77

19. Let us assume that the arc CD has been found (Figure 77). Rotate the segment BD about the center O of the circle S through an angle α; it will be taken into a new segment $B'C$ that makes an angle $ACB' = \alpha$ with the segment AC.

Thus we have the following construction: Rotate the point B about O through an angle α into a new position B'. Through the points A and B' pass a circular arc subtending an angle α (that is, if C is any point on the circular arc, then $\angle ACB' = \alpha$). The intersection of this circular arc with the circle S determines the point C.

The problem can have up to four solutions (the arc can meet the circle in two points, and the point B can be rotated about the point O in two directions).

20. Assume that the problem has been solved. Rotate the circle S_1 about A through an angle α into the position S_1' (Figure 78). The circles S_2 and S_1' will cut off equal chords on the line l_2. Thus the problem has

been reduced to Problem 8(c). In other words, a line l_2 must be passed through A so that it cuts off equal chords on S_1' and S_2. Then l_1 can be obtained by a rotation of l_2 about A through an angle α, and S_1 will cut the desired segment from l_1.

The problem can have up to four solutions. [Since S_1 can be rotated about A in either of two directions, there are two ways of reducing the problem to Problem 8(c) which, in turn may have two solutions.]

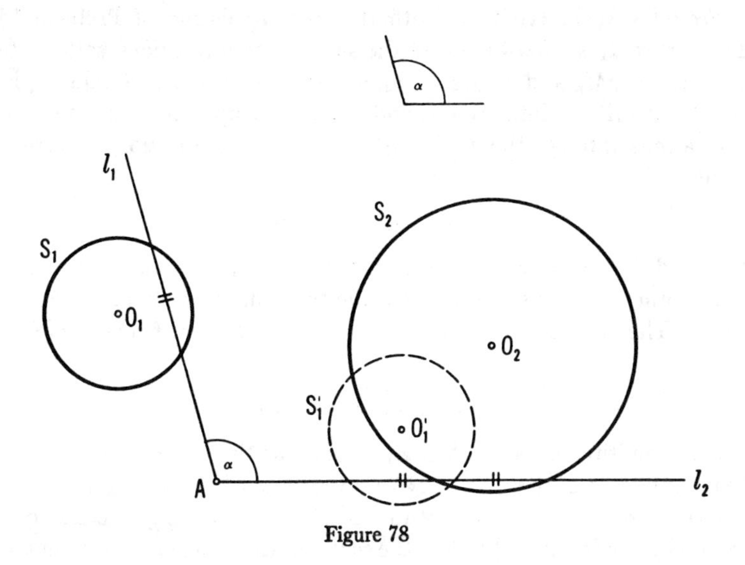

Figure 78

21. *First solution* (compare with the first solution of Problem 15). Assume that the problem has been solved and that $A_1A_2\cdots A_n$ is the desired n-gon (see Figure 79, where $n = 6$). Choose an arbitrary point B_1 in the plane. The sequence of rotations, first about M_1 through an angle α_1, then about M_2 through an angle α_2, etc., and finally about M_n through an angle α_n carries the segment A_1B_1 first into a segment A_2B_2, then carries A_2B_2 into a segment A_3B_3, $\cdots$ and finally carries A_nB_n into A_1B_{n+1}. All these segments are equal and therefore the vertex A_1 of the n-gon is equidistant from the points B_1 and B_{n+1} (where B_{n+1} is obtained from B_1 by these n rotations). Now choose a second point C_1 in the plane, and rotate it successively about the points $M_1, M_2, \cdots, M_n$ through angles $\alpha_1, \alpha_2, \cdots, \alpha_n$. Thus we obtain a second pair of points C_1 and C_{n+1} equidistant from A_1. Thus the vertex A_1 of the n-gon can be found as the intersection of the perpendicular bisectors of the segments B_1B_{n+1} and C_1C_{n+1}. Having found A_1 we obtain A_2 by rotating A_1 about M_1 through an angle α_1; A_3 is obtained by rotating A_2 about M_2 through an

angle α_2, etc. The problem has a unique solution provided that the perpendicular bisectors to B_1B_{n+1} and to C_1C_{n+1} do intersect (that is, the segments B_1B_{n+1} and C_1C_{n+1} are not parallel). If the perpendicular bisectors are parallel then the problem has no solution, and if they coincide then the problem has infinitely many solutions.

The polygon obtained as the solution to the problem need not be convex and may even intersect itself.

Second solution (compare with the second solution of Problem 15). The vertex A_1 is a fixed point of the sum of the n rotations with centers $M_1, M_2, \cdots, M_n$ and angles $\alpha_1, \alpha_2, \cdots, \alpha_n$ (these rotations take A_1 into A_2, A_2 into A_3, A_3 into A_4, etc. and, finally, A_n into A_1). But the sum of n rotations through the angles $\alpha_1, \alpha_2, \cdots, \alpha_n$ is a rotation through the angle

$$\alpha_1 + \alpha_2 + \cdots + \alpha_n,$$

provided that $\alpha_1 + \alpha_2 + \cdots + \alpha_n$ is not a multiple of 360°; it is a translation otherwise (this follows from the theorem on the sum of two rotations). The only fixed point of a rotation is the center of rotation. Therefore if

$$\alpha_1 + \alpha_2 + \cdots + \alpha_n$$

is not a multiple of 360°, then A_1 is found as the center of the rotation, that is, the sum of the rotations about the points $M_1, M_2, \cdots, M_n$ through angles $\alpha_1, \alpha_2, \cdots, \alpha_n$. Actually to find A_1 we may apply repeatedly the method given in the text to find the center of the sum of two rotations.†

A translation has no fixed points whatever. Therefore if

$$\alpha_1 + \alpha_2 + \cdots + \alpha_n$$

is a multiple of 360° then the problem has no solution in general. However, in the special case when the sum of the rotations about the points $M_1, M_2, \cdots, M_n$ through the angles $\alpha_1, \alpha_2, \cdots, \alpha_n$ (where the sum $\alpha_1 + \alpha_2 + \cdots + \alpha_n$ is a multiple of 360°) is the identity transformation, the problem has infinitely many solutions (any point in the plane may be chosen for the vertex A_1).

Thus, if $\alpha_1 = \alpha_2 = \cdots = \alpha_n = 180°$ (this is the case considered in Problem 15), the problem has a unique solution when n is odd and has no solution or has infinitely many solutions when n is even.

† It may happen that in the construction we shall have to find the center of a rotation that is the sum of a translation and a rotation. In this connection one should consult the text in fine print on page 36 or on page 51.

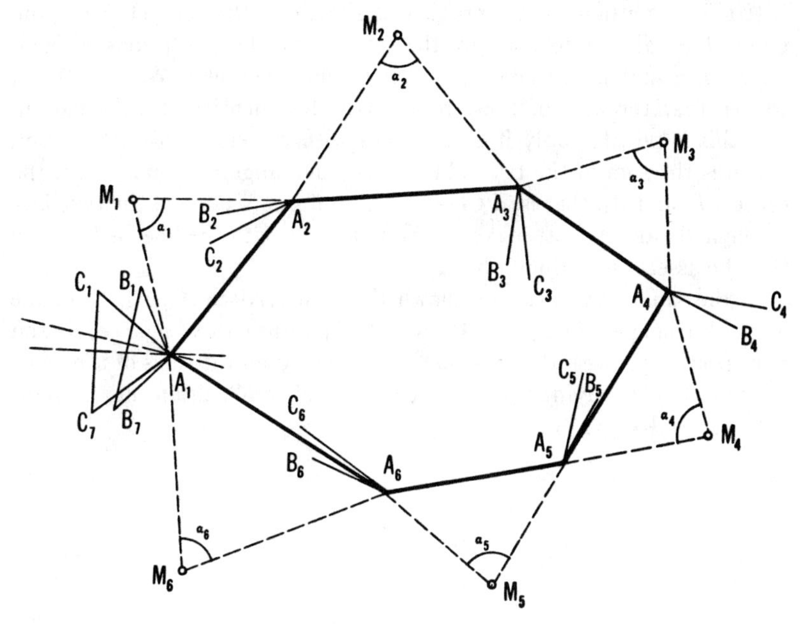

Figure 79

22. (a) Consider the sequence of three rotations, each through 120°, about the points O_1, O_2, O_3 (see Figure 31 in the text). The first of these rotations carries A into B, the second carries B into C, and the third carries C into A.

Thus the point A is a fixed point of the sum of these three rotations. But the sum of three rotations through 120° is, in general, a translation, and therefore has no fixed points. From the fact that A is a fixed point we see that the sum of these three rotations must be the identity transformation (translation through zero distance). The sum of the first two rotations is a rotation through 240° about the point O of intersection of two lines, one through O_1 and the other through O_2, each making an angle of 60° with O_1O_2. Therefore the triangle O_1O_2O is equilateral. Since the sum of this rotation and the rotation about O_3 through 120° is the identity transformation, the point O must coincide with O_3. Thus the triangle $O_1O_2O_3$ is equilateral, which was to be proved.

In the same way one can show that the centers O_1', O_2', O_3' of the equilateral triangles constructed on the sides of the given triangle ABC, but lying towards the interior of ABC, also form an equilateral triangle (Figure 80).

(b) The solution to this problem is similar to that of (a). Since the point A is taken into itself by the sum of the three rotations through angles β, α, and γ $(\alpha + \beta + \gamma = 360°)$ about the centers B_1, A_1, and C_1, we see that the sum of these rotations is the identity transformation. But this is possible only if C_1 coincides with the center of the rotation which is the sum of the two rotations through angles β and α about the centers B_1 and A_1, that is, if C_1 is the point of intersection of the two lines through B_1 and A_1 that make angles $\frac{1}{2}\beta$ and $\frac{1}{2}\alpha$ with the line B_1A_1. From this the assertion of the problem follows.

In the same way it can be shown that the vertices A_1', B_1', C_1' of the isosceles triangles ABC_1', BCA_1', and ACB_1' with vertex angles α, β, and γ, respectively, $(\alpha + \beta + \gamma = 360°)$ constructed on the sides of the given triangle ABC but lying towards the interior of ABC, also form a triangle with angles $\frac{1}{2}\alpha$, $\frac{1}{2}\beta$, $\frac{1}{2}\gamma$.

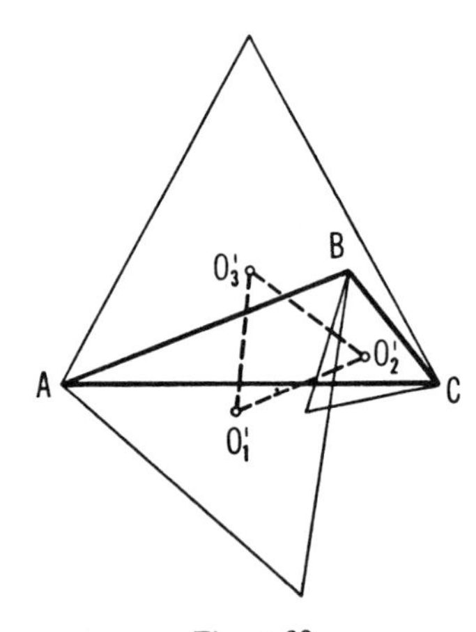

Figure 80

23. The sequence of three rotations in the same direction through angles of 60°, 60°, and 240° about the points A_1, B_1, and M takes the point B into itself (see Figure 32 in the text). Therefore the sum of these three rotations is the identity transformation, and thus the sum of the first two rotations is a rotation with center M. From this the assertion of the problem follows. (Compare with the solution to Problem 22.)

24. (a) The sum of the four rotations with centers M_1, M_2, M_3, and M_4, each through an angle of 60°, where the direction of the first and third rotations is opposite to that of the second and fourth, carries the

vertex A of the quadrilateral into itself (see Figure 33a, in the text). But the sum of the two rotations about M_1 and M_2 is a translation given by the segment M_1M_1', where M_1' is a vertex of the equilateral triangle $M_1M_2M_1'$ ($M_2M_1 = M_2M_1'$, $\measuredangle M_1M_2M_1' = 60°$, and the direction of rotation from M_2M_1 to M_2M_1' coincides with the direction of rotation from M_2B to M_2C; see Figure 81a, and Figure 28b in the text). Similarly the sum of the rotations about M_3 and M_4 is a translation given by the segment M_3M_3', where triangle $M_3M_4M_3'$ is equilateral (and the direction of rotation from M_4M_3 to M_4M_3' is the same as the direction of rotation from M_4D to M_4A). Thus the sum of two translations—given by the segments M_1M_1' and M_3M_3'—carries the point A into itself. But if the sum of two translations leaves even one point fixed, then this sum must be the identity transformation, that is, the two segments that determine the two translations must be equal, parallel, and oppositely directed. But if the equilateral triangles $M_1M_2M_1'$ and $M_3M_4M_3'$ are so situated that

$$M_1M_1' = M_3M_3', \qquad M_1M_1' \parallel M_3M_3'$$

and if M_1M_1' and M_3M_3' are oppositely directed, then the sides M_1M_2 and M_3M_4 are also equal, parallel, and oppositely directed, from which it follows that the quadrilateral $M_1M_2M_3M_4$ is a parallelogram (see Figure 81a).

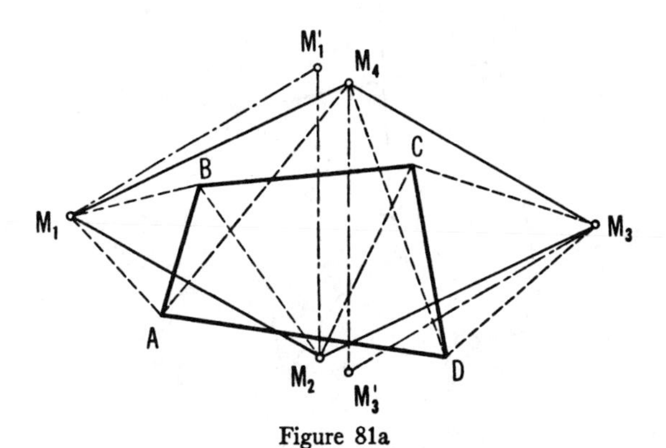

Figure 81a

(b) The sum of the four rotations about the points M_1, M_2, M_3, and M_4, each through an angle of 90°, clearly carries the vertex A of the quadrilateral into itself. It follows that this sum of four rotations is the identity transformation [compare the solution of Problem (a)]. But the sum of the rotations about M_1 and M_2 is a half turn about a point O_1—the vertex of an isosceles right triangle $O_1M_1M_2$ (since

$$\angle O_1M_1M_2 = \angle O_1M_2M_1 = 45°;$$

compare Figure 81b with Figure 28a in the text). Similarly the sum of rotations about M_3 and M_4 is a half turn about the vertex O_2 of an isosceles right triangle $O_2M_3M_4$. From the fact that the sum of the half turns about O_1 and O_2 is the identity transformation it clearly follows that these two points coincide. But this means that triangle $O_1M_1M_3$ is obtained from triangle $O_1M_2M_4$ by a rotation through 90° about the point $O_1 = O_2$, and therefore the segments M_1M_3 and M_2M_4 are equal and perpendicular.

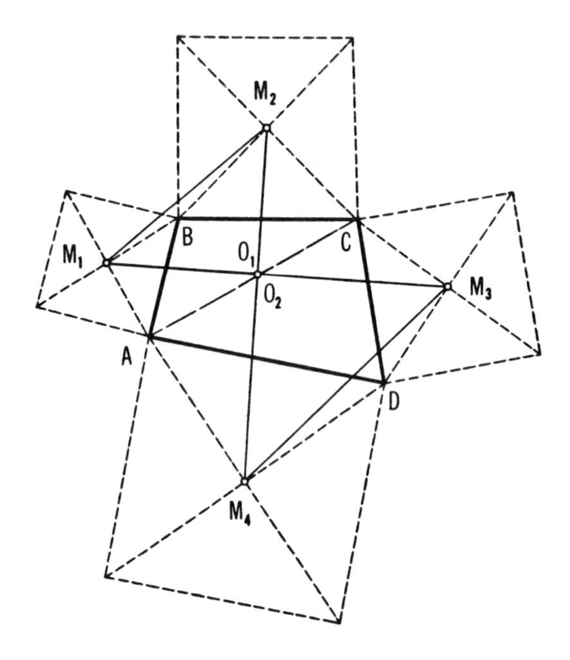

Figure 81b

(c) By what has already been proved [see the solution to Problem (b)], the diagonals M_1M_3 and M_2M_4 of the quadrilateral $M_1M_2M_3M_4$ are equal and mutually perpendicular. Further, since the point O of intersection of the diagonals of the parallelogram $ABCD$ is its center of symmetry, it is also the center of symmetry for all of Figure 81c, and in particular it is the center of symmetry for the quadrilateral $M_1M_2M_3M_4$ (which must, therefore, be a parallelogram—since the parallelogram is the only quadrilateral that has a center of symmetry). But a parallelogram whose diagonals are equal and perpendicular must be a square.

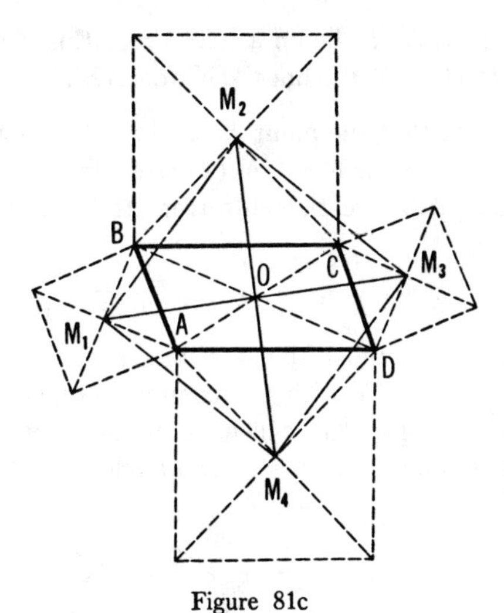

Figure 81c

In the same way it can be shown that if the four squares are constructed in the interior of the parallelogram, then their centers again form a square (Figure 81d).

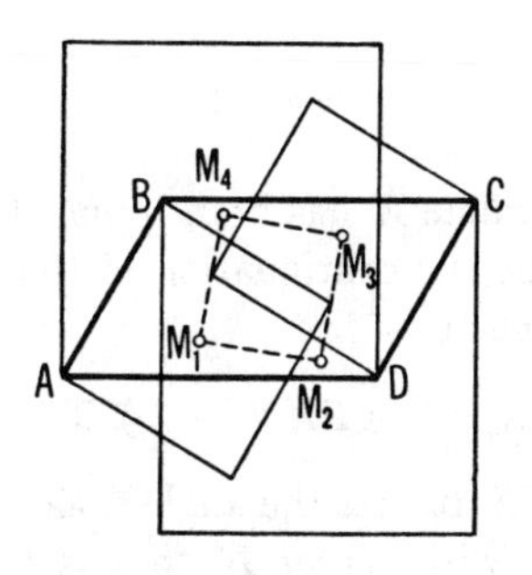

Figure 81d

Chapter Two. Symmetry

25. (a) Let us assume that the point X has been found, that is, that

$$\measuredangle AXM = \measuredangle BXN$$

(Figure 82a). Let B' be the image of B in the line MN; then

$$\measuredangle B'XN = \measuredangle BXN = \measuredangle AXM,$$

that is, the points A, X, B' lie on a line. From this it follows that X is the point of intersection of the lines MN and AB'.

(b) Let us assume that the point X has been found and let S_2' be the image of the circle S_2 in the line MN (Figure 82b).

If XA, XB, and XB' are tangents from the point X to the circles S_1, S_2, and S_2' then

$$\measuredangle B'XN = \measuredangle BXN = \measuredangle AXM,$$

that is, the points A, X, and B' lie on a line. Therefore X is the point of intersection of the line MN with the common tangent line AB' to the circles S_1 and S_2'. The problem can have at most four solutions (there are at most four common tangents to two circles).

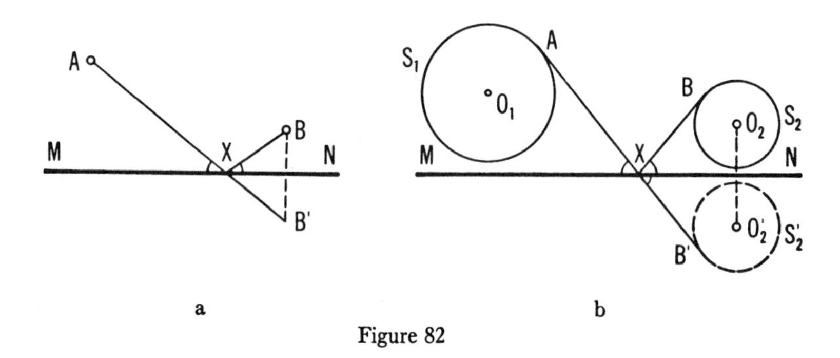

Figure 82

(c) *First solution.* Assume X has been found. Let B' be the image of B in MN and let XC be the continuation of the segment AX past the point X (Figure 83a). Then

$$\measuredangle CXN = 2\measuredangle BXN = 2\measuredangle B'XN,$$

and therefore the ray XB' bisects the angle NXC. Thus the line AXC is tangent to the circle S with center B' that is tangent to MN; consequently, the point X is the intersection of the line MN and the tangent from A to the circle S.

Second solution. Again, assume X has been found. Let A' be the image of A in the line $B'X$ (we are using the same notation as in the first solution). $B'X$ bisects the angle AXM; therefore A' lies on the line XM and $B'A = B'A'$ (Figure 83b). Thus A' can be found as the intersection of the line MN with the arc of a circle with center B' and radius $B'A$. The point X is now obtained as the intersection of the line MN with the perpendicular dropped from B' onto AA'.

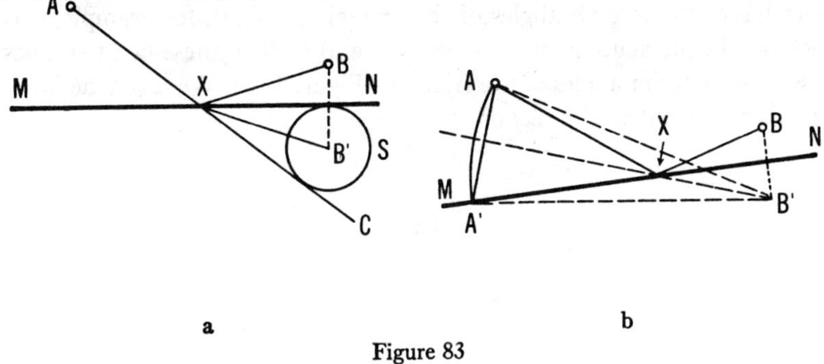

Figure 83

26. (a) Assume that the triangle ABC has been constructed, with l_2 bisecting angle B and l_3 bisecting angle C (Figure 84a). Then the lines BA and BC are images of each other in l_2, and the lines BC and AC are images of each other in l_3, and therefore the points A' and A'' obtained from A by reflection in the lines l_2 and l_3 lie on the line BC.

Thus we have the following construction: Reflect the point A in the lines l_2 and l_3 to obtain the points A' and A''. The vertices B and C are the points of intersection of the line $A'A''$ with the lines l_2 and l_3.

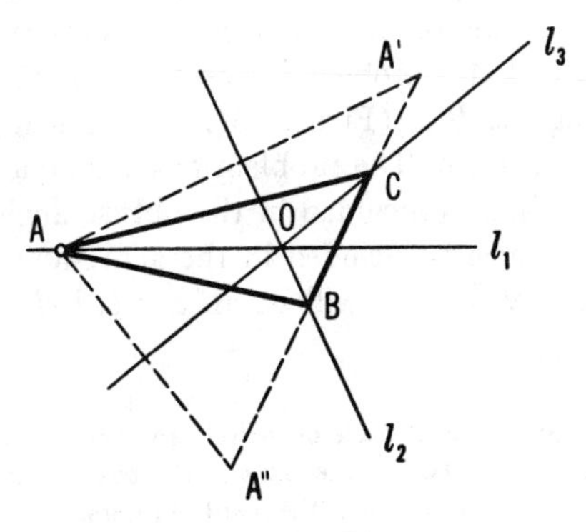

Figure 84a

If l_2 and l_3 are perpendicular, then the line $A'A''$ passes through the point of intersection of the three given lines and the problem has no solution; if l_1 is perpendicular to one of the lines l_2 and l_3, then $A'A''$ will be parallel to the other line and again the problem will have no solution. In case no two of the three given lines are perpendicular, the problem has a unique solution; however only in case each of the three given lines is included in the obtuse angle formed by the other two will the three lines bisect the *interior* angles of the triangle ABC; if, for example, l_1 is included in the acute angle formed by l_2 and l_3, then these last two lines bisect the exterior angles of the triangle (Figure 84b). We leave the proof of this statement to the reader.

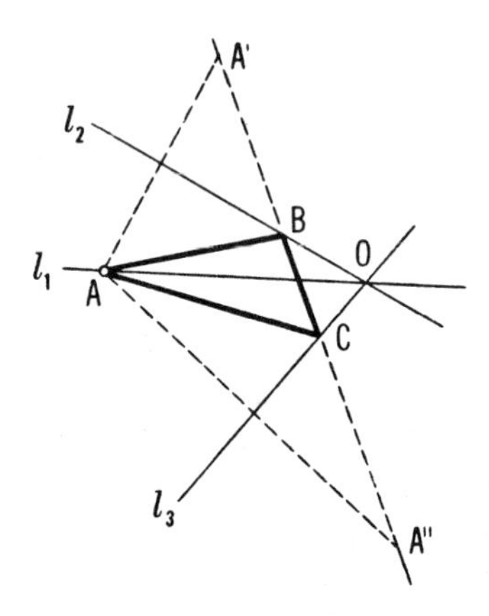

Figure 84b

(b) Choose an arbitrary point A' on one of the lines and construct the triangle $A'B'C'$ having the lines l_1, l_2, and l_3 as bisectors of its interior angles [see part (a) of this problem]. Construct tangents to S parallel to the sides of triangle $A'B'C'$ (Figure 85). The triangle thus obtained is the solution to the problem. The problem has a unique solution if each of the three lines l_1, l_2, l_3 is included in the obtuse angle formed by the other two; if one of them is included in the acute angle formed by the other two then the given circle will be an *escribed circle* or *excircle*[T] of the triangle.

[T] Every triangle has an inscribed circle or incircle and three excircles. Each excircle is tangent to the extensions of two of the sides of the triangle and to the third side (externally). The center of each excircle is the point of intersection of an internal angle bisector and the bisectors of the exterior angles at the other two vertices.

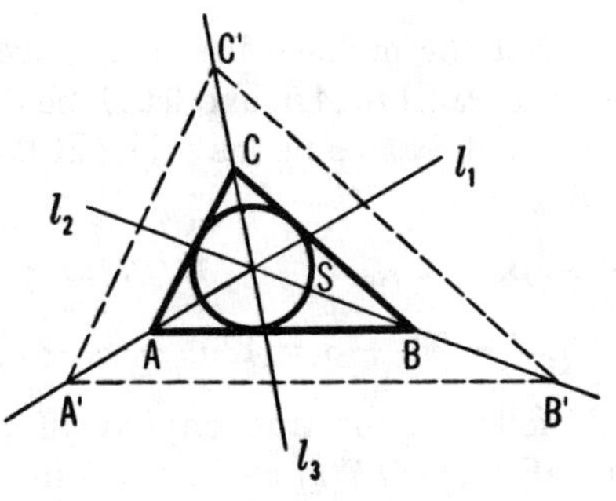

Figure 85

(c) Let us assume that the triangle ABC has been found (Figure 86). Since the point A is the image of the point B in the line l_2, it must lie on the line that is the image of BC in l_2; and since A is the image of C in l_3, it must also lie on the line that is the image of BC in l_3.

Thus we have the following construction: Pass a line m through A_1 perpendicular to l_1. Then construct the lines m' and m'' obtained from m by reflection in the lines l_2 and l_3. The point of intersection of m' and m'' will be the vertex A of the desired triangle; the vertices B and C are the images of this vertex in the lines l_2 and l_3 (Figure 86).

If the lines l_2 and l_3 are perpendicular, then either the lines m' and m'', obtained from m by reflection in l_2 and l_3, will be parallel (provided that the point A_1 does not coincide with the point O of intersection of the three lines l_1, l_2, and l_3) or they will coincide (if A_1 coincides with O). In the first case the problem has no solution, while in the second the solution is not determined uniquely. In all other cases the solution is unique.

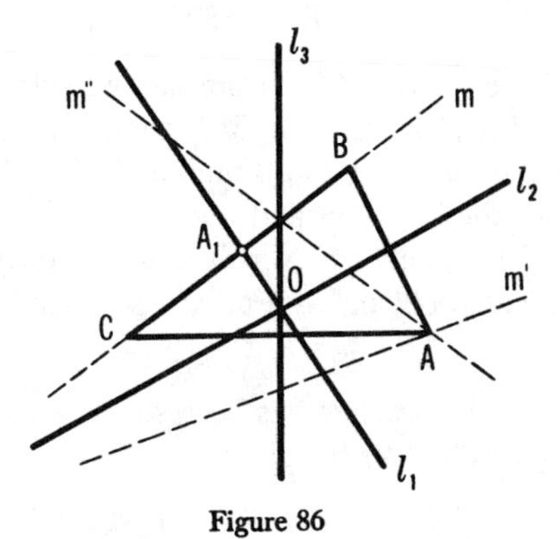

Figure 86

27. (a) Assume that the problem has been solved. Pass a line MN through the vertex C parallel to AB, and let B' be the image of B in the line MN (Figure 87). Let α and β be the angles at the base AB (we shall assume that $\alpha > \beta$). Then

$$\measuredangle ACN = 180° - \alpha, \qquad \measuredangle B'CN = \measuredangle BCN = \beta;$$

$$\measuredangle ACB' = (180° - \alpha) + \beta = 180° - (\alpha - \beta) = 180° - \gamma.$$

Thus we have the following construction: Lay off the segment $AB = a$, and construct a parallel line MN at a distance h from AB. Let B' be the image of B in the line MN. On the segment AB' construct the arc that subtends an angle of $180° - \gamma$. The point of intersection of this arc with the line MN is the vertex C of the triangle. The problem has a unique solution.

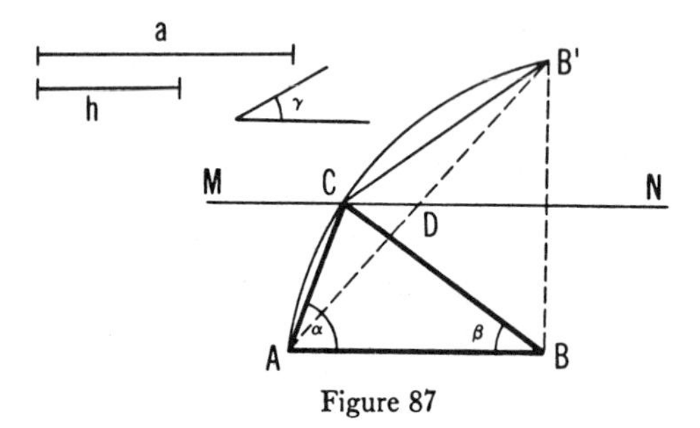

Figure 87

(b) Assume that the problem has been solved and determine the line MN and the point B' as in part (a) (Figure 87).

Since

$$\measuredangle ACB' = 180° - \gamma,$$

we can construct the triangle ACB' from the two sides AC and $CB' = BC$ and their included angle $180° - \gamma$. MN coincides with the median CD of this triangle (because MN is a "midline" of triangle ABB', that is, MN is parallel to the base AB and midway between this base and the opposite vertex B'). Finally, the vertex B is obtained as the image of B' in the line MN. The problem has a unique solution.

28. Assume that the problem has been solved and let B' be the image of B in OM (Figure 88). We have:

$$\measuredangle B'XA = \measuredangle B'XB + \measuredangle YXZ;$$

but

$$\measuredangle B'XB = 2\measuredangle OXZ = 2(\measuredangle XZY - \measuredangle MON)$$

(because $\measuredangle XZY$ is an exterior angle of triangle XOZ). Consequently

$$\begin{aligned}\measuredangle B'XA &= 2\measuredangle XZY - 2\measuredangle MON + \measuredangle YXZ \\ &= \measuredangle XZY + \measuredangle XYZ + \measuredangle YXZ - 2\measuredangle MON \\ &= 180° - 2\measuredangle MON.\end{aligned}$$

Thus $\measuredangle B'XA$ is known. Now X can be found as the point of intersection of the ray OM with the arc, constructed on the chord AB', that subtends an angle equal to $180° - 2\measuredangle MON$. The problem has a unique solution.

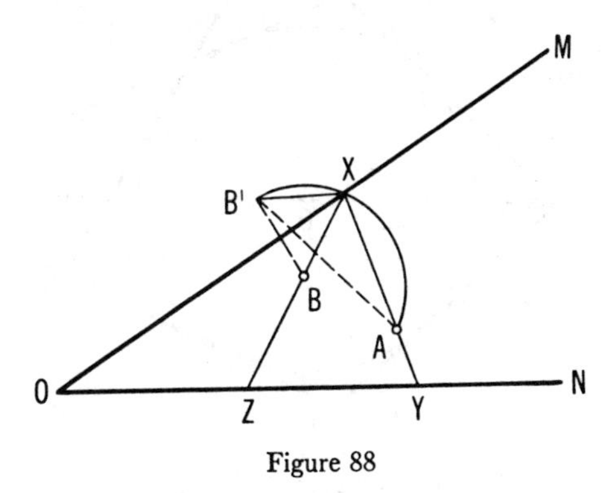

Figure 88

29. (a) Assume that the quadrilateral $ABCD$ has been constructed and let B' be the image of B in the diagonal AC (Figure 89). Since $\measuredangle BAC = \measuredangle DAC$ the point B' lies on the line AD. The three sides of the triangle $B'DC$ are known:

$$DC, \quad B'C = BC, \quad \text{and} \quad DB' = AD - AB' = AD - AB.$$

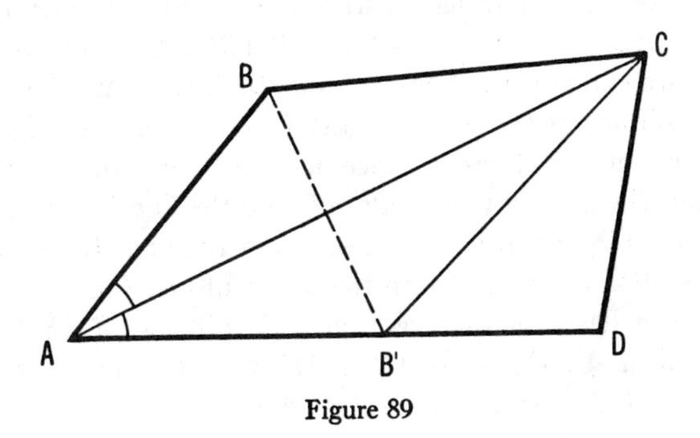

Figure 89

Construct this triangle, and locate the vertex A (this can be done since the distance AD is known). The vertex B is then obtained as the image of B' in the line AC. The problem has a unique solution if $AD \neq AB$; it has no solution whatsoever if $AD = AB$ and $CD \neq CB$; it has more than one solution if $AD = AB$ and $CD = CB$.

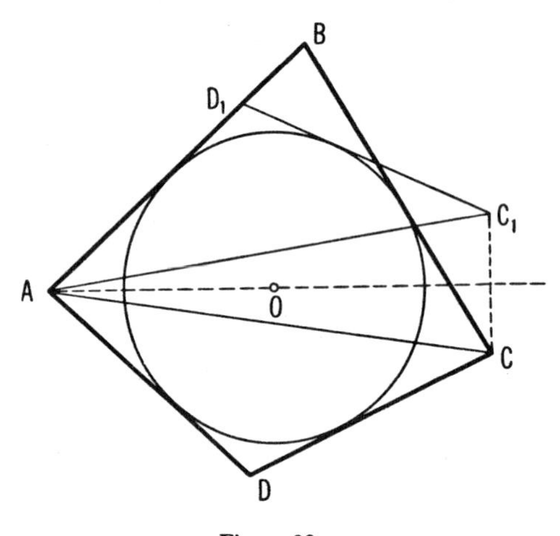

Figure 90

(b) Assume that the problem has been solved (Figure 90), and let triangle AD_1C_1 be the image of triangle ADC in the line AO (O is the center of the circle inscribed in the quadrilateral). Clearly the point D_1 lies on the line AB, and the side D_1C_1 is tangent to the circle inscribed in the quadrilateral $ABCD$.

Thus we have the following construction: On an arbitrary line lay off the segments AB and $AD_1 = AD$. Since $\measuredangle ABC$ and $\measuredangle AD_1C_1 = \measuredangle ADC$ are known, we can find the lines BC and D_1C_1 (although we do not yet know the positions of the points C and C_1 on these lines). Now we can construct the inscribed circle since it is tangent to the three lines AB, BC, and D_1C_1. Finally, the side AD and the line DC are obtained as the images of AD_1 and D_1C_1 by reflection in the line AO. (The point C is the intersection of line BC with the image of line D_1C_1.)

The problem has a unique solution if $\measuredangle ADC \neq \measuredangle ABC$; it has no solution at all if $\measuredangle ADC = \measuredangle ABC$, $AD \neq AB$; it has more than one solution if $\measuredangle ADC = \measuredangle ABC$, $AD = AB$.

30. (a) Assume that the problem has been solved, that is, that points $X_1, X_2, \cdots, X_n$ have been found on the lines $l_1, l_2, \cdots, l_n$ such that

$$AX_1X_2\cdots X_nB$$

is the path of a billiard ball (in Figure 91 the case $n = 3$ is represented). It is easy to see that the point X_n is the point of intersection of the line l_n with the line $X_{n-1}B_n$, where B_n is the image of B in l_n [see the solution to Problem 25(a)], that is, the points B_n, X_n, X_{n-1} lie on a line. But then the point X_{n-1} is the point of intersection of the line l_{n-1} with the line $X_{n-2}B_{n-1}$, where B_{n-1} is the image of B_n in l_{n-1}. Similarly one shows that the point X_{n-2} is the intersection of the lines l_{n-2} and $X_{n-3}B_{n-2}$, where B_{n-2} is the image of B_{n-1} in l_{n-2}; the point X_{n-3} is the intersection of the lines l_{n-3} and $X_{n-4}B_{n-3}$, where B_{n-3} is the image of B_{n-2} in l_{n-3}, and so forth.

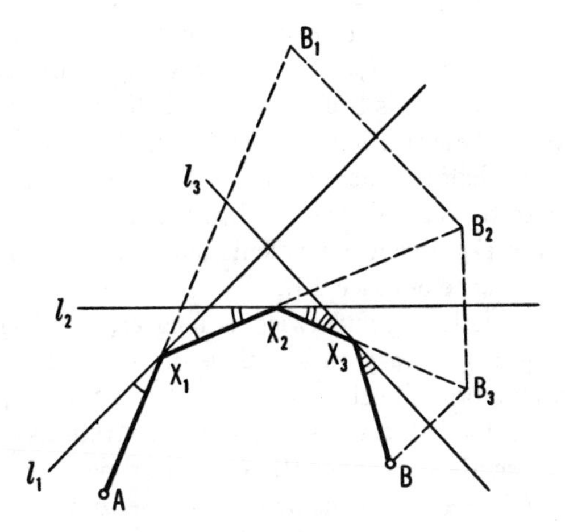

Figure 91

Thus we have the following construction: Reflect the point B in l_n, obtaining the point B_n; next reflect B_n in l_{n-1} to obtain B_{n-1}, and so forth, until the image B_1 of the point B_2 in line l_1 is obtained. The point X_1, that determines the direction in which the billiard ball at A must be hit, is obtained as the point of intersection of the line l_1 with the line AB_1. It is then easy to find the points $X_2, X_3, \cdots, X_n$ with the aid of the points $B_2, B_3, \cdots, B_n$ and X_1.

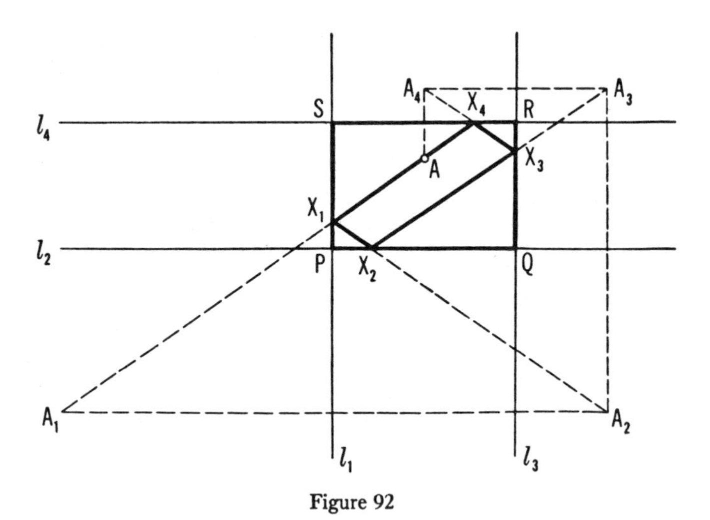

Figure 92

(b)[T] Following the procedure of part (a), we first reflect the point A in l_4 to obtain A_4, then reflect A_4 in l_3 to obtain A_3, and so forth until we reach A_1 (see Figure 92). It is easily verified that reflection in l_4 followed by reflection in l_3 is equivalent to a half turn about the point of intersection, R, of these two lines.[TT] Similarly, the next two reflections are equivalent to a half turn about the point P. Hence the four reflections are equivalent to the sum of two half turns, about R and P. But as we know (see Figure 17), this is equivalent to a translation in the direction PR through a distance of twice PR.

Thus AA_1 is parallel to, and twice as long as, the diagonal PR. By considering angles it is easy to see that the path $AX_1X_2X_3X_4A$ is a parallelogram (the opposite sides are parallel) with sides parallel to the diagonals. Thus if the ball is not stopped when it returns to the point A, it will describe exactly the same path a second time.

Finally, it can be seen from the figure that the total length of the path is equal to AA_1, that is, to twice the length of a diagonal.

31. (a) Let us assume that the problem has been solved. Draw the circle S_1 of center A and radius a, and the circle S_2 of center X and radius XB (Figure 93a). Clearly these two circles are tangent at a point lying on the line AX. Since S_2 passes through the point B, it must also

[T] This solution was inserted by the translator in place of the original solution.

[TT] See page 50.

pass through the point B', the image of B in the line l. Thus the problem has been reduced to the construction of a circle S_2, passing through two known points B and B' and tangent to a given circle S_1, that is, to Problem 49(b) of Vol. 2.[T] The center X of the circle S_2 is the desired point. This problem has at most two solutions; there may only be one or there may be none at all.

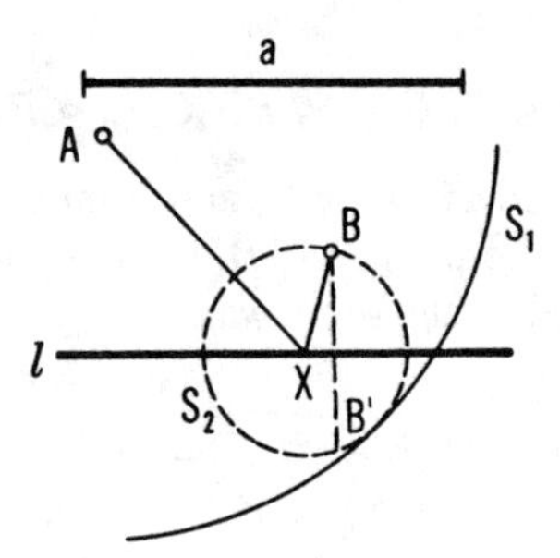

Figure 93a

(b) Assume that the problem has been solved, let S_1 be the circle of center A and radius a, and let S_2 be the circle of center X and radius BX (Figure 93b). The circles S_1 and S_2 are tangent at a point that lies on the line AX. In addition S_2 passes through the point B' that is the image of B in the line l. Therefore this problem is also reduced to Problem 49(b) of Vol. 2.[T] There are at most two solutions.

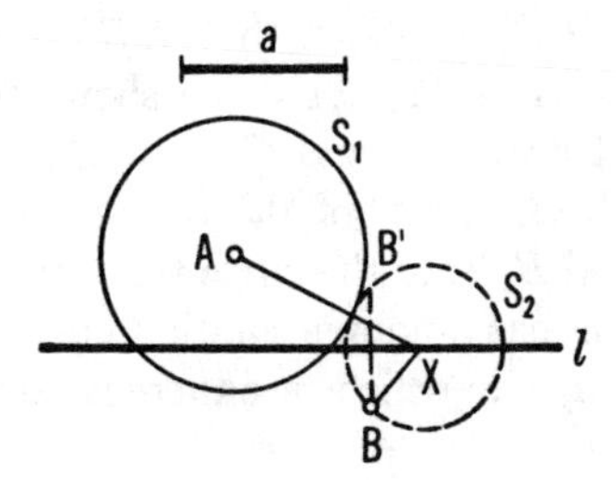

Figure 93b

[T] Since at this time Volume 2 is not available in English, we refer the reader to p. 175, Problem V of *College Geometry* by Nathan Altschiller-Court, Johnson Publishing Co., 1925, Richmond.

32. (a) Let H_1 be the image of H in the side BC (Figure 94). Let P, Q, R be the feet of the altitudes. We have

$$\measuredangle BH_1C = \measuredangle BHC \qquad \text{(because} \qquad \triangle BH_1C \cong \triangle BHC).$$

But

$$\measuredangle BHC = \measuredangle RHQ,$$

and

$$\measuredangle RHQ + \measuredangle RAQ = \measuredangle BH_1C + \measuredangle RAQ = 180°;$$

therefore $\measuredangle BH_1C + \measuredangle BAC = 180°$, and from this it follows that H_1 lies on the circle through the points A, B, C. The images of H in sides AB and AC can be treated in the same way.

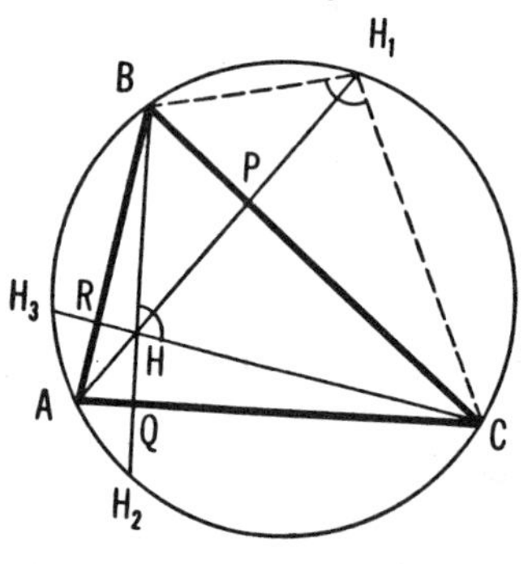

Figure 94

(b) Let us assume that triangle ABC has been constructed. The points H_1, H_2, and H_3 lie on the circumscribed circle [see Problem (a)]. Since

$$\measuredangle BRC = \measuredangle BQC (= 90°)$$

and $\measuredangle BHR = \measuredangle CHQ$, it follows that $\measuredangle RBH = \measuredangle QCH$, that is, arc AH_3 is equal to arc AH_2. Similarly one shows that arcs BH_1 and BH_3 are equal, and that arcs CH_1 and CH_2 are equal. From this it follows that the vertices A, B, and C of the triangle are the midpoints of the arcs H_2H_3, H_3H_1, and H_1H_2 of the circle through the three points H_1, H_2, and H_3. The problem has a unique solution unless the points H_1, H_2, and H_3 lie on a straight line, in which case there is no solution at all.

33. (a) Clearly; for example, the altitudes of triangle $A_2A_3A_4$ are the lines

$$A_1A_4 \perp A_2A_3, \qquad A_1A_3 \perp A_2A_4, \qquad \text{and} \qquad A_1A_2 \perp A_3A_4;$$

the point of intersection of these altitudes is the point A_1.

(b) Let A_4' be the image of A_4 in the line A_2A_3 (Figure 95). This point lies on the circle S_4, circumscribed about triangle $A_1A_2A_3$ [see Problem 32(a)]. Thus the circle circumscribed about triangle $A_2A_4'A_3$ coincides with S_4; from this it follows that the circle S_1, circumscribed about triangle $A_2A_3A_4$, is congruent to S_4 (S_1 and S_4 are images of each other in the line A_2A_3). Similarly one shows that the circles S_2 and S_3 are also congruent to S_4.

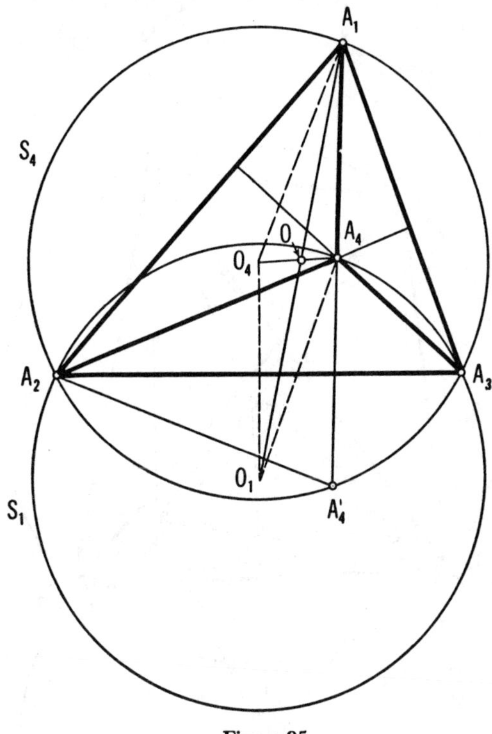

Figure 95

(c) At least one of the triangles $A_1A_2A_3$, $A_1A_2A_4$, $A_1A_3A_4$, and $A_2A_3A_4$ must be acute angled; indeed, if triangle $A_2A_3A_4$ has an obtuse angle at A_4, then triangle $A_2A_3A_1$ (where A_1 is the point of intersection of the altitudes of triangle $A_2A_3A_4$) will be acute. Thus we shall assume that triangle $A_1A_2A_3$ is acute and that the point A_4 lies inside it.

Consider the quadrilateral $A_1A_4O_1O_4$. The points O_1 and O_4 are centers of circles S_1 and S_4 that are images of each other in the line A_2A_3 [see Figure 95 and the solution to part (b) of this problem]. Therefore O_1 and O_4 are images of each other in A_2A_3, and so $O_1O_4 \perp A_2A_3$. In the quadrilateral $A_1A_4O_1O_4$ we thus have

$$O_4O_1 \parallel A_1A_4 \quad \text{and} \quad O_1A_4 = O_4A_1 = R$$

(where R is the radius of the circles S_1, S_2, S_3, and S_4). Therefore this quadrilateral is either a parallelogram or an isosceles trapezoid. But it cannot be an isosceles trapezoid because the perpendicular bisector A_2A_3 of side O_4O_1 does not meet side A_1A_4. Hence $A_1A_4O_1O_4$ is a parallelogram and its diagonals A_1O_1, A_4O_4 meet in a point O that is the midpoint of each of them. In the same way one shows that O is the midpoint of A_2O_2 and of A_3O_3.

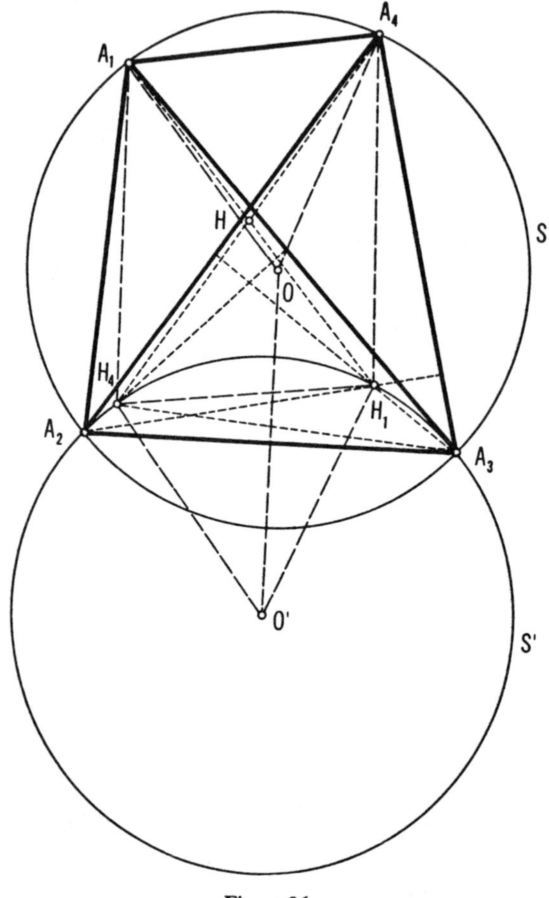

Figure 96

34. (a) Let O' be the image of the center O of the circle S in the line A_2A_3 (Figure 96). The quadrilaterals $OO'H_4A_1$ and $OO'H_1A_4$ are parallelograms [see the solution to Problem 33(c)]. Therefore

$$A_1H_4 = OO' = A_4H_1, \qquad A_1H_4 \parallel OO' \parallel A_4H_1,$$

and so $A_1H_4H_1A_4$ is a parallelogram. From this it follows that the segments A_1H_1 and A_4H_4 have a common midpoint H. In the same way one shows that H is also the midpoint of A_2H_2 and A_3H_3.

(b) By comparing Figure 96 and Figure 95 one sees that, for example, H_4 lies on the circle S', the image of S in the line A_2A_3; H_1 also lies on this circle. Thus A_2, A_3, H_1, and H_4 all lie on a circle congruent to S. The remaining assertions of the theorem are proved similarly.

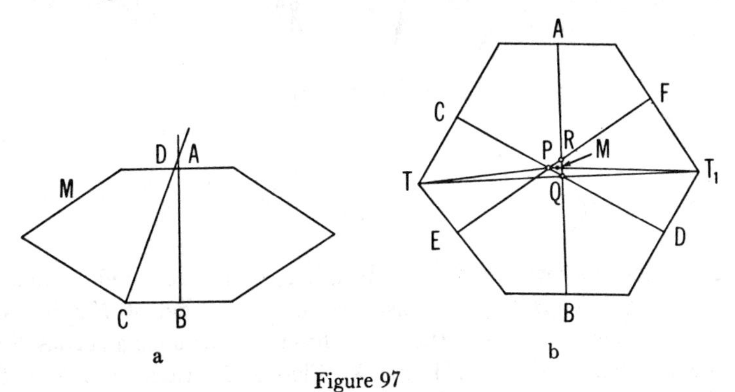

Figure 97

35. First of all it is clear that any two axes of symmetry AB and CD of the polygon M must intersect inside M; indeed, if this were not the case (Figure 97a), then they could not both divide the figure into two parts of equal area. Now let us show that if there is a third axis of symmetry EF, then it must pass through the point of intersection of the first two. Assume that this were not the case; then the three axes of symmetry AB, CD, and EF would form a triangle PQR (Figure 97b). Let M be a point inside this triangle. It is easy to see that each point in the plane lies on the same side of at least one of these three axes of symmetry as does M. Let T be the vertex of the polygon that is farthest from M (if there is more than one such vertex, let T be any one of them), and let T and M lie on the same side of the axis of symmetry AB. Thus, if T_1 is the image of T in AB (T_1 is therefore a vertex of the polygon), then $MT_1 > MT$ (since the projection of MT_1 onto TT_1 is larger than the projection of MT on TT_1; see Figure 97b). This contradiction proves the theorem.

[In a similar way it can be shown that if any bounded figure (not necessarily a polygon) has several axes of symmetry, then they must all pass through a common point. For unbounded figures this is not so: Thus, the strip between two parallel lines l_1 and l_2 has infinitely many axes of symmetry, perpendicular to l_1 and l_2 and all parallel to each other.]

Remark: The assertion of this problem is evident from mechanical considerations. The center of gravity of a homogeneous, polygonal-shaped body, having an axis of symmetry, must lie on that axis. Consequently, if there are several axes of symmetry they must all pass through the center of gravity.

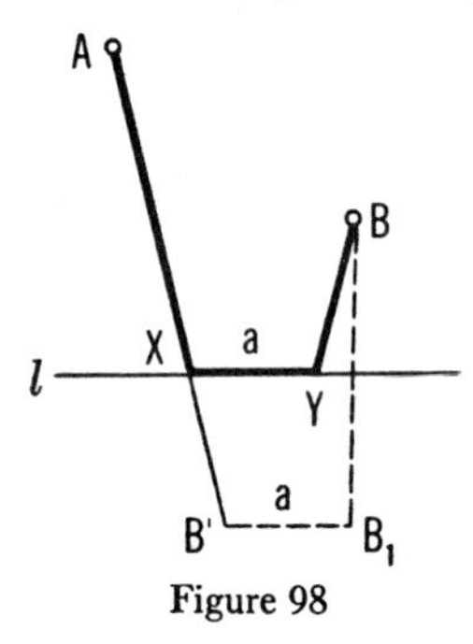

Figure 98

36. Since the segment XY has length a, we are required to minimize the sum $AX + BY$. Let us assume that the segment XY has been found. A glide reflection in the axis l through a distance a carries B into a new point B', and carries Y into X (Figure 98); therefore $BY = B'X$, and so

$$AX + BY = AX + B'X.$$

Thus it is required that the path AXB' should have minimum length. From this it follows that X is the point of intersection of l with AB'.

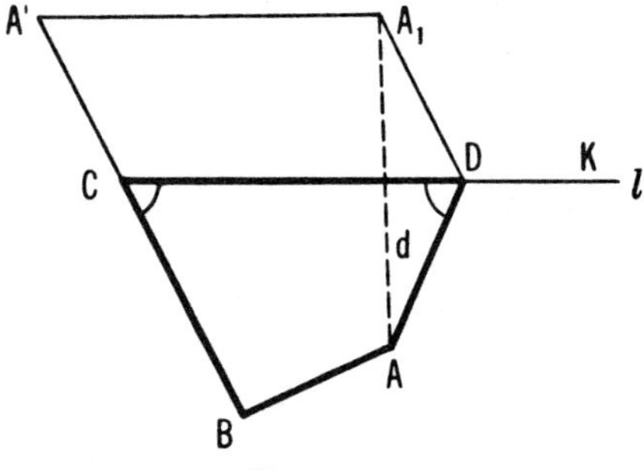

Figure 99

37. (a) Assume that the quadrilateral $ABCD$ has been constructed. Let A' be the image of A under a glide reflection in the axis DC through a distance DC (Figure 99); then $\sphericalangle A'CD = \sphericalangle ADK$ (where DK is the extension of side DC past the point D) because if A_1 is the image of A in DC, then

$$\sphericalangle A'CD = \sphericalangle A_1DK = \sphericalangle ADK.$$

But

$$\sphericalangle ADK = 180° - \sphericalangle D = 180° - \sphericalangle C;$$

consequently, $\sphericalangle A'CD = 180° - \sphericalangle C$, that is, $A'CB$ is a straight line. In addition we know that

$$A'B = A'C + CB = AD + CB,$$

and we know the distance d from A to CD.

Thus we have the following construction: Let l be any line, let A be a point at a distance d from l, and let A' be the image of A under a glide reflection in the line l through a distance CD. The vertex B of the quadrilateral can now be found, since we know the distances AB and

$$A'B = AD + BC.$$

The vertex C is the point of intersection of the segment $A'B$ with the line l, and the vertex D which lies on l is found by laying off the known distance CD from the point C. The problem can have two, one, or no solutions.

(b) Draw the segment AB; the line l can now be found as the common tangent to the two circles of radii d_1 and d_2, with centers at the points A and B respectively (Figure 100). It remains to put the segment DC on the line l in such a position that the sum of the lengths $AD + BC$ has the given value [compare with Problem 31(a)].

Assume that the points D and C have been found and let A' and A'' be the images of A under a translation in the direction of the line l through a distance DC, and under a glide reflection with axis l through a distance DC. Clearly the circle of center C and radius AD passes through the points A' and A'' $(A'C = A''C = AD)$ and is tangent to the circle S with center B and radius

$$BC + CA'' = BC + AD.$$

But the circle S can be constructed from the given data, and thus it only remains to find the circle passing through the two known points A'

and A'' and tangent to S [see Problem 49(b) of Vol. 2].[T] The center of this circle is the vertex C.

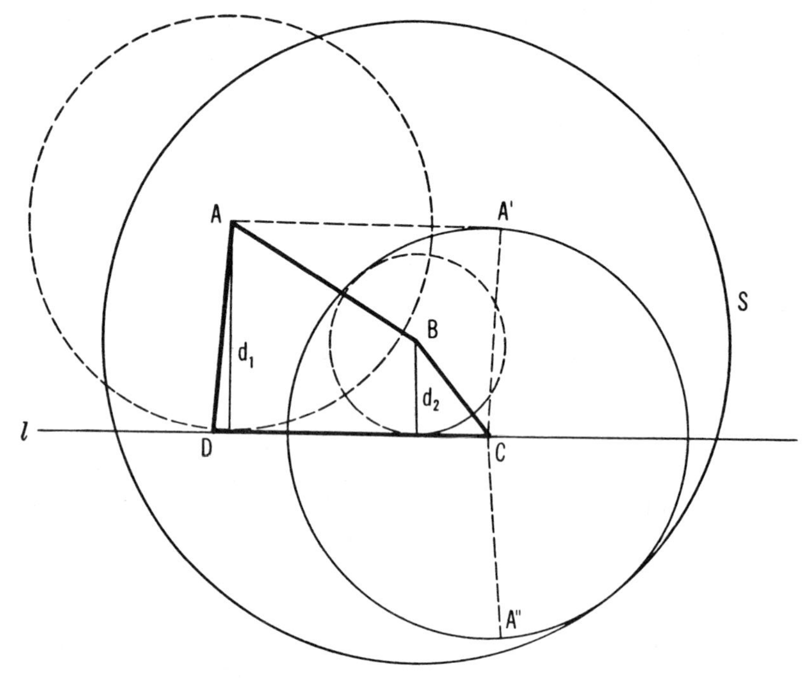

Figure 100

38. *First solution.* Clearly the only time a ray of light will be reflected from a mirror in a direction exactly opposite to the direction of incidence is when the path is perpendicular to the mirror. From now on we shall assume that the ray of light does not strike the first side of the angle at right angles. Let us now consider the case when the ray MN, after two reflections in the angle ABC, leaves along a path PQ exactly opposite to MN (Figure 101a). In this case we have:

$$\angle PNB + \angle NPB = 180° - \angle NBP = 180° - \alpha;$$

$$\begin{aligned} 2(180° - \alpha) &= 2\angle PNB + 2\angle NPB \\ &= \angle ANM + \angle PNB + \angle NPB + \angle CPQ \\ &= 180° - \angle MNP + 180° - \angle NPQ \\ &= 360° - (\angle MNP + \angle NPQ). \end{aligned}$$

[T] See translator's note on page 107.

Since the rays MN and PQ are parallel and oppositely directed,

$$\measuredangle MNP + \measuredangle NPQ = 180°,$$

so

$$2(180° - \alpha) = 360° - 180°, \qquad \text{and} \qquad \alpha = 90°.$$

Conversely, if $\alpha = 90°$ then $\measuredangle MNP + \measuredangle QPN = 180°$, that is, the direction of the departing ray PQ is opposite to MN.

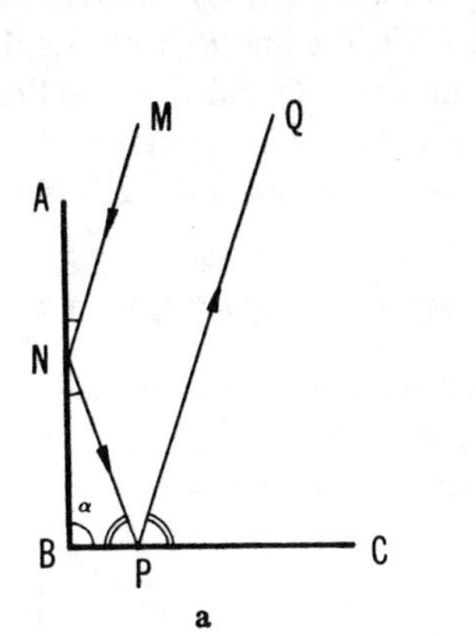

a b

Figure 101

Next consider the case when the incoming ray MN, after four reflections in the sides of the angle, leaves in a direction RS opposite to MN (Figure 101b; the only way in which a light ray can leave in the opposite direction to the direction of incidence after exactly three reflections is if it hits the second side of the angle at right angles; this cannot happen for every incoming light ray—in fact, for a given angle α there is only one angle of incidence for which this will happen). Reflect the line AB and the path PQR in the line BC; the line BA_1 is the image of BA, and the point Q_1 is the image of Q in BC. Then

$$\measuredangle ABA_1 = 2\measuredangle ABC = 2\alpha.$$

Further

$$\measuredangle QPB = \measuredangle Q_1PB = \measuredangle NPC;$$

therefore, NPQ_1 is a straight line. In the same way it can be shown that Q_1RS is a straight line (since $\measuredangle QRB = \measuredangle Q_1RB = \measuredangle SRC$). Finally, $\measuredangle BQ_1P = \measuredangle A_1Q_1R$, since these angles are equal respectively to the angles BQP and AQR, which are equal. Thus we see that the ray MN, reflected from the points N and Q_1 of the angle $ABA_1 = 2\alpha$, leaves in a direction Q_1S, opposite to the incoming direction. But then by what was shown previously $2\alpha = 90°$ and therefore $\alpha = 90°/2$. Conversely, if $\alpha = 90°/2$ then $\measuredangle ABA_1 = 90°$ and so the ray MN, after four reflections in the sides of the angle ABC, leaves in the opposite direction to the direction of incidence.

Now consider the case when the incoming ray MN is reflected six times in the sides of the angle, and then leaves along a path TU opposite to the incoming path (Figure 101c; in general a light ray cannot leave along a path opposite to the incoming path after exactly five reflections). Reflect the line AB and the path $PQRST$ in the line BC; let BA_1 be the image of BA and let Q_1 and S_1 be the images of Q and S in the line BC. Just as before we can conclude that NPQ_1 is a straight line ($\measuredangle Q_1PB = \measuredangle QPB = \measuredangle NPC$), that S_1TX is a straight line ($\measuredangle S_1TB = \measuredangle STB = \measuredangle UTC$) and that

$$\measuredangle Q_1RB = \measuredangle S_1RC, \qquad \measuredangle RQ_1B = \measuredangle PQ_1A_1, \qquad \measuredangle RS_1B = \measuredangle TS_1A_1.$$

Thus we find that the ray MN, reflected successively from the lines AB, BA_1, BC, and again from BA_1 at the points N, Q_1, R, and S_1 leaves in the direction S_1U, opposite to the incoming direction MN.

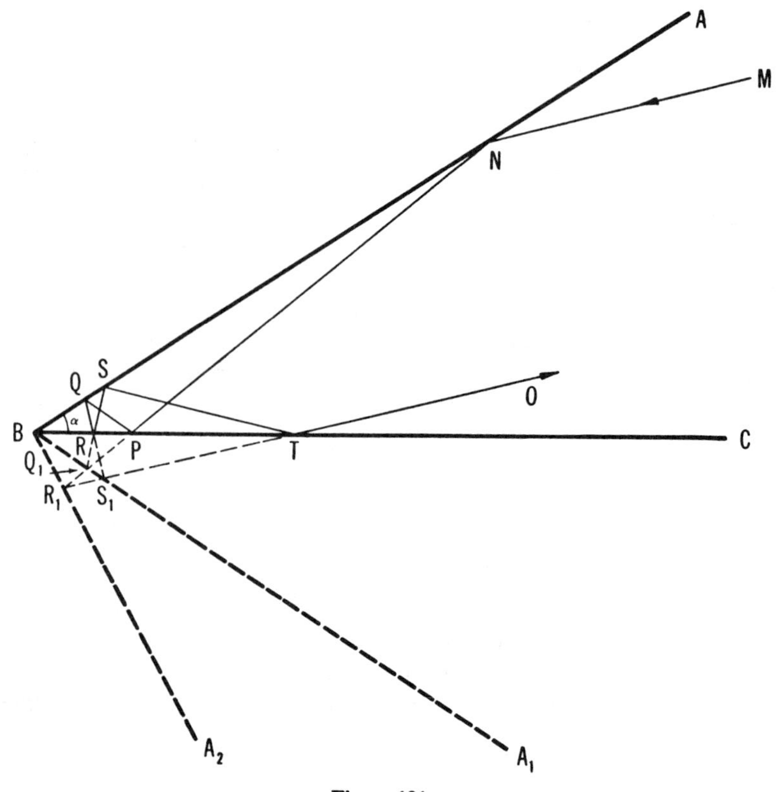

Figure 101c

Now reflect the line BC and the path Q_1RS_1 in the line BA_1; let BA_2 be the image of BC and let R_1 be the image of R in the line BA_1. Then NPQ_1R_1 is a straight line (because $\sphericalangle R_1Q_1B = \sphericalangle RQ_1B = \sphericalangle PQ_1A_1$) and R_1S_1TU is a straight line (because $\sphericalangle R_1S_1B = \sphericalangle RS_1B = \sphericalangle TS_1A_1$), and $\sphericalangle Q_1R_1B = \sphericalangle S_1R_1A_2$ (because they are equal respectively to the angles Q_1RB and S_1RC, which are equal). Thus we find that the ray MN after being reflected in the sides of the angle ABA_2 ($= 3\alpha$) at the points N and R_1 leaves in the direction R_1U, opposite to the incoming direction MN. But then by what was proved earlier we must have $3\alpha = 90°$, that is, $\alpha = 90°/3$. Conversely, if $\alpha = 90°/3$, then $\sphericalangle ABA_2 = 90°$ and the ray MN, after being reflected six times in the sides of angle ABC, leaves in the direction opposite to the direction of incidence.

Finally, suppose that after $2n$ reflections in the sides of an angle $ABC = \alpha$ the ray leaves in the direction opposite to the direction of the incoming ray [in general a light ray cannot leave in a direction opposite to the direction of incidence after $(2n - 1)$ réflections in the sides of an angle].

Proceed as in the previous cases,[T] that is, if the incoming ray strikes AB, reflect the path of the ray in line BC; let BA_1 be the image of AB after this reflection. Next, reflect BC in line BA_1 to obtain BA_2, then reflect BA_1 in BA_2 to obtain BA_3, and so forth, until, after $n - 1$ reflections, we have BA_{n-1}. The angle $ABA_{n-1} = n\alpha$.

Next, establish that the incoming ray, when continued by the proper reflections, forms a straight line which hits $A_{n-1}B$, is reflected there, then hits BA so that it leaves in the direction opposite to that of its entry. Then, by what was proved earlier, conclude that $n\alpha = 90°$, and hence, that

$$\alpha = \frac{90°}{n}.$$

Second solution. Let ABC be the given angle, and let $MNPQ\cdots$ be the path of the light ray (see Figure 102a, where the case $n = 2$, $\alpha = 45°$ is shown). We are only interested in the directions of the path, and it will be convenient to have all these directions emanate from a single point O (in the figure

$$O1 \parallel MN, \qquad O2 \parallel NP, \qquad O3 \parallel PQ,$$

and so forth). Since $\sphericalangle MNA = \sphericalangle PNB$, it follows that the ray $O2$ is the image of $O1$ in the line $OU \parallel AB$ (to prove this it is sufficient to note

[T] In the Russian version of this book, the details of this proof were carried out. We have omitted them here in order to save space and to avoid the somewhat complicated notation.

that in Figure 102a, NM' is the image of NP in NB). Similarly, the ray $O3$ is the image of $O2$ in the line $OV \parallel BC$. Therefore by Proposition 3 on page 50, the ray $O3$ is obtained from the ray $O1$ by a rotation through an angle $2\sphericalangle UOV = 2\alpha$. Similarly the ray $O5$ is obtained from the ray $O3$ by a rotation through an angle 2α in the same direction; consequently the ray $O5$ is obtained from the ray $O1$ by a rotation through an angle 4α, and so forth. Therefore, if $\alpha = 90°/n$ then the ray $O(2n + 1)$, which has the same direction as that of a light ray after n reflections from each of the two faces of the angle, will form an angle $n \cdot 2\alpha = 180°$ with the ray $O1$, which establishes the assertion of the problem. [Here we are assuming that $0 < \sphericalangle MNA < \alpha$; if $\sphericalangle MNA > \alpha$, then MN will intersect BC, which means that the incoming light ray has to be reflected from side BC before it can hit side BA. This fact guarantees that the rays in the directions $O1$, $O3$, $O5$, $\cdots$, etc. will all hit the mirror BA, while the rays in the directions $O2$, $O4$, $\cdots$, etc. will hit the mirror BC. If $\sphericalangle MNA = \alpha$, that is, if the incoming ray MN is parallel to side BC, then the ray $O(2n)$ will already be opposite in direction to $O1$: In this case the final ray leaves along a path opposite to the path of the original incoming ray; however the number of reflections is one fewer than in the general case; see Figure 102b, where $\sphericalangle ABC = 45°$, $\sphericalangle MNA = 45°$.]

These considerations show that if $\alpha \neq 90°/n$, then not every incoming light ray will, after successive reflections in the sides, leave in a direction opposite to the direction of approach of the original ray.

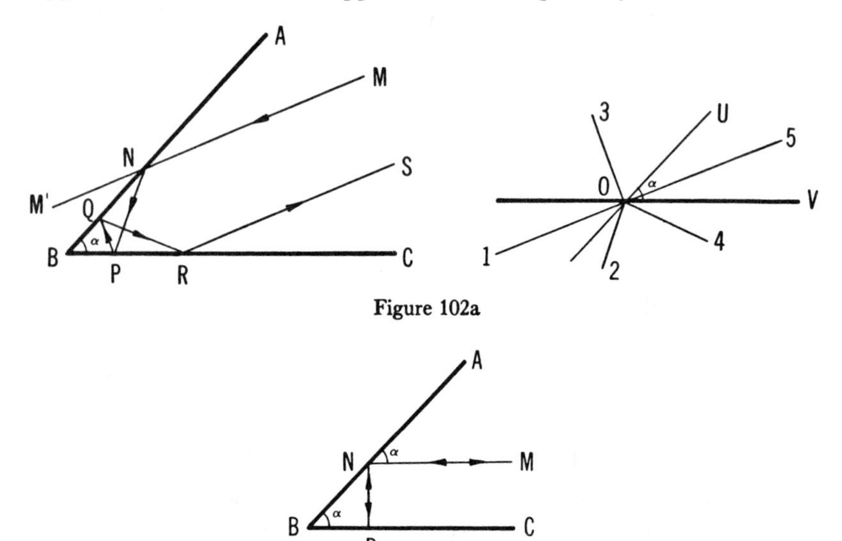

Figure 102a

Figure 102b

39. (a) *First solution* (see also the first solutions to Problems 15 and 21). Let $A_1, A_2, \cdots, A_n$ be the desired n-gon and let B_1 be any point in the plane. Reflect the segment A_1B_1 successively in the lines

$$l_1, \ l_2, \ \cdots, \ l_{n-1}, \ l_n;$$

we obtain segments $A_2B_2, A_3B_3, \cdots, A_nB_n, A_1B_{n+1}$. Since these segments are all congruent to each other, it follows that $A_1B_1 = A_1B_{n+1}$, that is, the point A_1 is equidistant from B_1 and B_{n+1}, and lies therefore on the perpendicular bisector of the segment B_1B_{n+1}.

Now choose another point C_1 in the plane and let $C_2, C_3, \cdots, C_n, C_{n+1}$ be the points obtained, starting from C_1, by successive reflections in the lines $l_1, l_2, \cdots, l_{n-1}, l_n$. Clearly the vertex A_1 of the n-gon is also equidistant from C_1 and C_{n+1}, and therefore lies on the perpendicular bisector to C_1C_{n+1}. Therefore A_1 can be found as the intersection of the perpendicular bisectors to the segments B_1B_{n+1} and C_1C_{n+1} (the segments B_1B_{n+1} and C_1C_{n+1} can be constructed, once we have chosen any two distinct points for B_1 and C_1). By reflecting A_1 successively in the n given lines we obtain the remaining vertices of the n-gon.

The problem has a unique solution provided that the segments B_1B_{n+1} and C_1C_{n+1} are not parallel (i.e., provided that the perpendicular bisectors p and q intersect in one point); if $B_1B_{n+1} \parallel C_1C_{n+1}$ then the problem has no solution when p and q are distinct, and has infinitely many solutions (the problem is undetermined) when p and q coincide.

The n-gon obtained as the solution to the problem may intersect itself.

One drawback to this solution is that it gives no indication of the essential difference between the cases when n is even and when n is odd (see the second solution to the problem).

Second solution (see also the second solutions to Problems 15 and 21). Let $A_1A_2 \cdots A_n$ be the desired n-gon (see Figure 50a). If we reflect the vertex A_1 successively in the lines $l_1, l_2, \cdots, l_{n-1}, l_n$ we obtain the points $A_2, A_3, \cdots, A_n$ and, finally, A_1 again. Thus, A_1 is a *fixed point* of the sum of the reflections in the lines $l_1, l_2, \cdots, l_n$.

We now consider separately two cases.

First case: n even. In this case the sum of the reflections in the lines $l_1, l_2, \cdots, l_n$ is, in general, a rotation about some point O (see page 55), which can be found by the construction used in the addition of reflections. The point O is the only fixed point of the rotation, and so A_1 must coincide with O. Having found A_1, one has no difficulty in finding all the remaining vertices of the n-gon. The problem has a unique solution in this case.

In the exceptional case, when the sum of the reflections in the lines $l_1, l_2, \cdots, l_n$ is a translation or is the identity transformation (a rotation through an angle of zero degrees, or a translation through zero distance), the problem either has no solution at all (a translation has no fixed points) or has more than one solution—any point in the plane can be taken for the vertex A_1 (every point is a fixed point of the identity transformation).

Second case: n odd. In this case the sum of the reflections in the lines $l_1, l_2, \cdots, l_n$ will, in general, be a glide reflection (see pages 55–56). Since a glide reflection has no fixed points, there will in general be no solution when n is odd. In the exceptional case, when the sum of the reflections in the lines $l_1, l_2, \cdots, l_n$ is a reflection in a line l (this line can be constructed), the solution will not be uniquely determined; any point of the line l can be taken for the vertex A_1 of the n-gon (every point of the axis of symmetry is a fixed point under reflection in this line).

(Thus, for $n = 3$, the problem has, in general, no solutions; the only exceptions are the cases when the lines l_1, l_2, l_3 meet in one point [see Problem 26(c)] or are parallel; in these cases the problem has more than one solution [see Proposition 4 on page 53]).

(b) This problem is similar to Problem (a). If $A_1A_2 \cdots A_n$ is the desired n-gon (see Figure 50b), then the line A_nA_1 is taken by successive reflections in the lines $l_1, l_2, \cdots, l_{n-1}, l_n$ into the lines

$$A_1A_2, \quad A_2A_3, \quad \cdots, \quad A_{n-1}A_n$$

and finally back into A_nA_1. Thus A_nA_1 is a *fixed line* of the sum of the reflections in the lines $l_1, l_2, \cdots, l_n$. We consider two cases.

First case: n even. In this case the sum of the reflections in the lines $l_1, l_2, \cdots, l_n$ is, in general, a reflection about some point O and, therefore, has in general no fixed lines. Thus for n even our problem has, in general, no solution. In the exceptional cases when the sum of the reflections is a half turn about the point O (a rotation through an angle of 180°), or is a translation, or is the identity transformation, the problem has more than one solution. In the first case one can take any line through the center of symmetry to be the line A_nA_1; in the second case one can take any line parallel to the direction of translation; in the third case one can take any line whatsoever in the plane.

Second case: n odd. In this case the sum of the reflections in the lines $l_1, l_2, \cdots, l_n$ is, in general, a glide reflection with an axis l (that can be constructed). Since l is the only fixed line of a glide reflection, it follows that the side A_nA_1 of the desired n-gon must lie on l; by reflecting l successively in the lines $l_1, l_2, \cdots, l_{n-1}$, we obtain all the remaining sides of the n-gon. Thus for odd n the problem has, in general, a unique solution. An exception occurs when the sum of the reflections in the given lines is a reflection in a line l; in this case the problem has more than one solution. For the side A_nA_1 one can take the line l itself, or any line perpendicular to it.

(Thus, for $n = 3$, the problem has in general a unique solution; the lines l_1, l_2, l_3 will either all be bisectors of the exterior angles of the triangle, or two of them will bisect interior angles and the third will bisect the exterior angle. The only exception is when the three lines l_1, l_2, and l_3 meet in a point; in this case the problem has more than one solution [see Problem 26(a)]; the lines l_1, l_2, l_3 will all bisect interior angles, or two or them will bisect exterior angles and the third will bisect the interior angle.)

We leave it to the reader to find a solution to part (b) similar to the first solution to part (a).

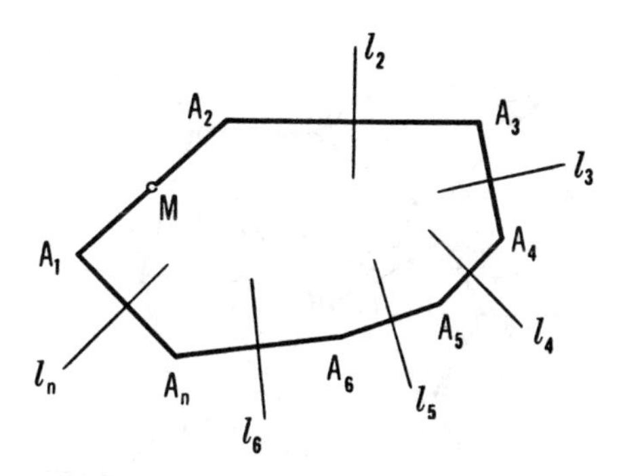

Figure 103

40. (a) Assume that the problem has been solved (Figure 103). A half turn about the point M will carry the vertex A_1 into A_2, a reflection in the line l_2 will carry the vertex A_2 into A_3, a reflection in l_3 will carry A_3 into A_4, and so forth. Finally, a reflection in l_n carries A_n into A_1. Thus, A_1 is a fixed point of the sum of a half turn about M followed by reflections in the lines $l_2, l_3, \cdots, l_n$. A half turn about the point M is equivalent to a pair of reflections in lines. We shall consider separately two cases.

First case: n odd. In this case the problem reduces to finding fixed points of the sum of an even number of reflections in lines. This sum is, in general, a rotation about some point O (which can be constructed from the point M and the lines $l_2, l_3, \cdots, l_n$). Therefore for odd n the problem has in general a unique solution [compare this with the first case in the solution to Problem 39(a)]. The only exceptional cases are when the sum of the even number of reflections in the lines is a translation—then the problem has no solution at all; or is the identity transformation—then the problem has many solutions.

Second case: n even. In this case the problem reduces to finding the fixed points of an odd number of reflections in lines. In general this sum is a glide reflection and the problem has no solution (a glide reflection has no fixed points). In the special case when the sum of the reflections is itself a reflection in some line l, the problem will have many solutions (reflection in a line has an infinite number of fixed points, namely all the points on the line l).

The construction can also be carried out in a similar manner to the construction in the first solution to Problem 39(a). The polygon obtained as the solution may intersect itself.

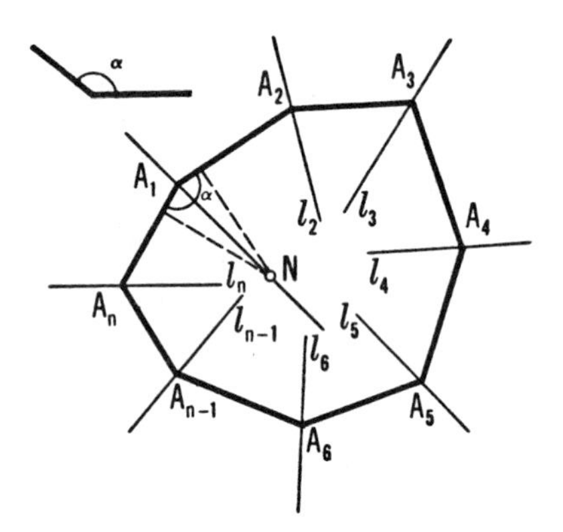

Figure 104

(b) Assume that the problem has been solved (Figure 104). A rotation of $180° - \alpha$ about the point M carries the line A_nA_1 into A_1A_2. A reflection in l_2 carries A_1A_2 into A_2A_3, a reflection in l_3 carries A_2A_3 into A_3A_4, and so forth. Finally, a reflection in l_n carries $A_{n-1}A_n$ into A_nA_1. Thus, A_nA_1 is a fixed line of the transformation consisting of the sum of a rotation through $180° - \alpha$ about the point M (which can be replaced by two reflections in lines) and $n - 1$ reflections in the lines $l_2, l_3, \cdots, l_n$.

We consider separately two cases.

First case: n even. The sum of an odd number of reflections in lines is in general a glide reflection; it has a unique fixed line, the axis of symmetry l (that can be constructed), and therefore the problem has a unique solution. In the special case when the sum of the reflections is a reflection in some line, the problem will have infinitely many solutions (because reflection in a line has infinitely many fixed lines).

Second case: n odd. In this case the transformation we are considering will be the sum of an even number of reflections in lines which, in general, is a rotation. In this case the problem will have no solution. In special cases, however, this sum of reflections may be a half turn about some point, a translation, or the identity transformation; in each of these cases the problem will have more than one solution.

The polygon that was constructed to solve the problem may intersect itself; the lines $l_2, l_3, \cdots, l_n$ will bisect either the exterior or the interior angles.

The construction can also be carried out in a manner similar to that in the first solution to Problem 39(a).

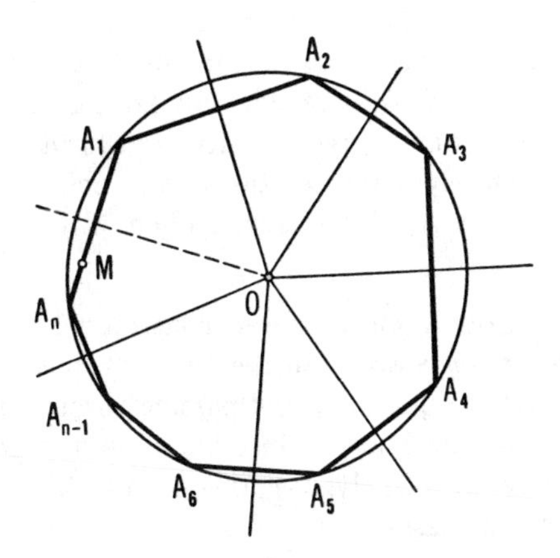

Figure 105

41. (a) Let $A_1A_2A_3\cdots A_n$ be the desired n-gon (Figure 105). Reflect the vertex A_1 successively in lines drawn from the center O of the circle and perpendicular to the sides A_1A_2, A_2A_3, $\cdots$, $A_{n-1}A_n$, A_nA_1 of the n-gon (these lines are known, since we are given the directions of the sides of the n-gon); the vertex A_1 is first taken into A_2, then A_2 is taken into A_3, $\cdots$, then A_{n-1} is taken into A_n, and finally A_n is taken back into A_1. Thus A_1 is a fixed point of the sum of n reflections in known lines. Let us consider two cases separately.

First case: n odd. Since the sum of three reflections in lines meeting in a point is again a reflection in some line through this point (See Proposition 4 on page 53), it is not difficult to see that the sum of any odd number of reflections in lines that all pass through a common point is again a reflection in some line through this point. (First replace the first three reflections by a single reflection, then consider the sum of this reflection and the next two, etc.) Therefore the sum of our n reflections is a reflection in some line passing through the center O of the circle. There are exactly two points on the circle that are left fixed by reflection in l—they are the points of intersection of the circle with l. Taking one of these points for the vertex A_1 of the desired polygon, we find the other vertices by successive reflections of this one in the n lines. The problem has two solutions.

Second case: n even. The sum of any two reflections in lines passing through the point O is a rotation about O through some angle. From this it follows that the sum of an even number, n, of reflections in lines passing through O may be replaced by the sum of $\frac{1}{2}n$ rotations about O; from this it is clear that the sum is itself a rotation about O. Since a rotation about O has, in general, no fixed points on a circle with center O, our problem has no solutions in general. An exception is the case when the sum of the n reflections is the identity transformation; in this case the problem has infinitely many solutions—any point on the circle can be chosen for the vertex A_1 of the desired n-gon.

(b) Assume that the n-gon has been constructed (see Figure 105). Reflect the vertex A_1 successively in the $(n - 1)$ lines perpendicular to the sides A_1A_2, A_2A_3, $\cdots$, $A_{n-1}A_n$ and passing through the center O of the circle (these lines are known, since we know the point O and the directions of the sides of the polygon); this process takes A_1 into A_n. We consider separately two cases.

First case: n odd. In this case the sum of $(n - 1)$ reflections in lines passing through the point O is a rotation about O through an angle α (that can be found). Thus, angle $A_1OA_n = \alpha$ is a known angle, and so we know the length of the chord A_1A_n and its distance to the center. Since A_1A_n must pass through a given point M, it only remains to pass tangents from the point M to the circle with center O and radius equal to the distance from the chord A_1A_n to the center O. The problem can have two, one, or no solutions.

Second case: n even. In this case the sum of $(n - 1)$ reflections in lines passing through a common point is a reflection in some line l through this point. Therefore A_1 and A_n are images of each other in l. Since A_1A_n must pass through a known point M, it can be found by simply dropping the perpendicular from M onto l. The problem always has a unique solution.

42. (a) Since the sum of the reflections in the three lines l_1, l_2, and l_3 meeting in the point O is a reflection in some line l (also passing through the point O), it follows that the point A_3 is obtained from A by a reflection in l. But A_6 is obtained from A_3 by a reflection in l, and so A_6 coincides with A.

This result is valid for any *odd* number of lines meeting in a point (compare Problem 13). If we have an even number n of lines meeting in a point O, then the sum of the n reflections in these lines is a rotation about O through some angle α, and so the point A_{2n} obtained after $2n$ rotations will coincide with the original point A only in case α is a multiple of 180°.

Remark. The point A_6 obtained from an arbitrary point A of the plane by six successive reflections in lines $l_1, l_2, l_3, l_1, l_2, l_3$ will coincide with the initial point A if and only if l_1, l_2, and l_3 meet in a point or are parallel [if $l_1 \parallel l_2 \parallel l_3$, then the sum of the reflections in l_1, l_2, and l_3 is a reflection in some line l, and the reasoning used in the solution to Problem 42(a) can be applied]. In all other cases the sum of the reflections in l_1, l_2, and l_3 is a glide reflection, and thus the point A_6 is obtained from A by two successive glide reflections along some axis l, that is, by a translation in the direction of l; therefore A_6 cannot coincide with A. [The sum of two (identical) glide reflections along an axis l can be written as the sum of the following four transformations: translation along l, reflection in l, reflection in l, and translation along l (see page 48), that is, as the sum of two (identical) translations along l.]

(b) This problem is essentially the same as part (a) [see also Problem 14(b)].

(c) The sum of the reflections in l_1 and l_2 is a rotation about their point of intersection O through some angle α; the sum of the reflections in l_3 and l_4 is a rotation about O through some angle β. From this it follows that (no matter in which order these reflections are performed!) the point A_4 is obtained from A by a rotation about O through an angle of $\alpha + \beta$, which was to be proved [compare with Problem 14(a)].

43. (a) Since the three lines CM, AN, BP meet in a point, it follows that the sum of the reflections in the lines CM, AN, BP, CM, AN, BP is the identity transformation [see Problem 42(a)]. To show that the lines CM', AN', BP' meet in a point it is sufficient to show that the sum of the reflections in the lines CM', AN', BP', CM', AN', BP' is also the identity transformation [see the remark following the solution to Problem 42(a)]. However reflection in the line CM' is the same as the sum of the reflections in the three lines CB, CM and CA all meeting in the point C—this follows from the fact that rotation through angle BCM' about the point C carries line CM into CA, and carries CB into line CM', which is the image of CM in the bisector of angle BCA (compare Figure 106a with Figure 47b, and see the proof of the second half of Proposition 4, page 53). Similarly, reflection in AN' is the same as the sum of the reflections in the three lines AC, AN, and AB, and reflection in BP' is the sum of the reflections in the lines BA, BP, and BC. From this it follows that the sum of the reflections in CM', AN', and BP' is the same as the sum of the reflections in the following nine lines: CB, CM, CA, $AC(= CA)$, AN, AB, $BA(= AB)$, BP, and BC. Since two consecutive reflections in the same line cancel each other, this is the same as the sum of the reflections in the following five lines: CB, CM, AN, BP, and BC. Now perform this transformation twice; we obtain the sum of the reflections in the following ten lines: CB, CM, AN, BP, BC, CB $(= BC)$, CM, AN, BP, and BC, which is the same as the sum of the reflections in the eight lines CB, CM, AN, BP, CM, AN, BP, and BC. But if the sum of the reflections in the six "inner" lines is the identity transformation, then the sum of our eight reflections in the eight lines reduces to the sum of the two reflections in CB and BC $(= CB)$, that is, to the identity transformation!

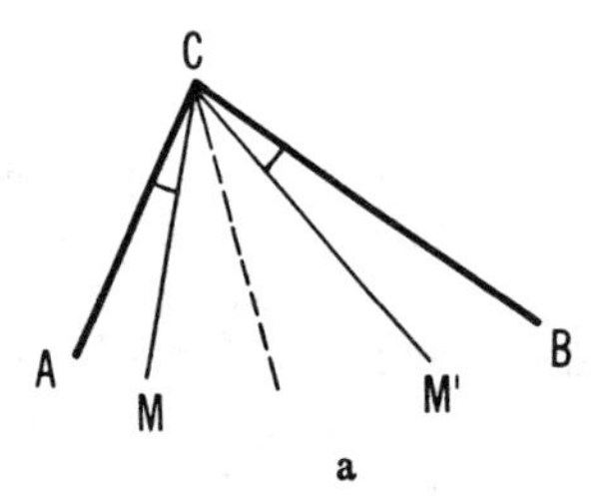

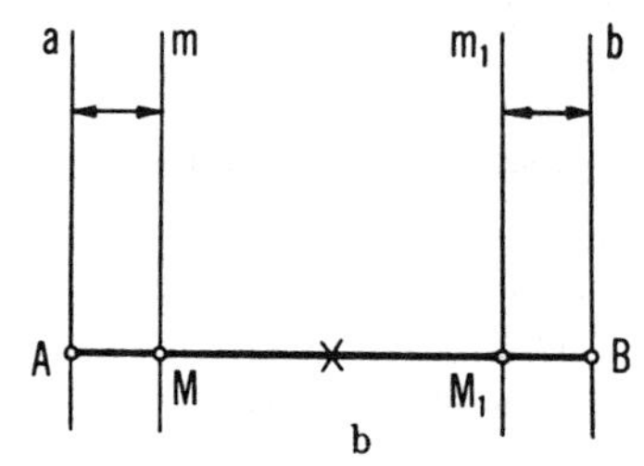

Figure 106

(b) Let the perpendiculars to the sides AB, BC and CA of the triangle ABC, erected at the points M and M_1, N and N_1, P and P_1 be denoted by m and m_1, n and n_1, p and p_1; let a and b denote the perpendiculars to side AB erected at the endpoints A and B. We must show that *if the sum of the reflections in the lines* m, n, p, m, n, p *is the identity transformation, then the sum of the reflections in the lines* $m_1, n_1, p_1, m_1, n_1, p_1$ *is also the identity transformation* [compare the solution to Problem (a)]; clearly the perpendiculars to two different sides of a triangle cannot be parallel to one another. But the reflection in m_1 is identical with the sum of the reflections in the point A, in the line m and in the point B; similarly, the reflection in n_1 is the sum of the reflections in B, n and C, and the reflection in p_1 is the sum of the reflections in C, p and A. To prove the first of these assertions, note that the reflection in A is the sum of the reflections in AB and a, and the reflection in B is the sum of the reflections in b and AB; thus, the sum of the reflections in A, m and B is equal to the sum of the reflections in the following five lines: AB, a, m, b, and AB. But the sum of the three "inner" reflections is equal to the reflection in m_1 alone—this follows from the fact that the translation of the two lines a and m, carrying m into b, carries a into m_1 (since m_1 is the reflection of m in the midpoint of the segment AB; compare Figure 106b with Figure 47a). Therefore the sum of the five reflections is equivalent to the sum of the reflections in the three lines: AB, m_1, and AB, or to the sum of the reflections in M_1 and AB. The reflection in M_1 is also equal to the sum of the reflections in m_1 and AB taken in that order; therefore the sum of the reflections in M_1 and AB is equal to the sum of the reflections in m_1, AB, and AB, and this is clearly the same as a single reflection in m_1 alone.

It is now clear that the sum of the reflections in the six lines $m_1, n_1, p_1, m_1, n_1, p_1$ is equal to the sum of the reflections in the following points and lines: $A, m, B; B, n, C; C, p, A; A, m, B; B, n, C; C, p, A$, or, what is the same thing, to the reflections in A, m, n, p, m, n, p, A. Therefore, if the sum of the six "inner" reflections is the identity transformation, then the sum of all the reflections (which reduces in this case to two reflections in the point A) is also the identity transformation [compare the solution to part (a)].

44. If we take the sum of the reflections in three lines in the plane twice, then we obtain either the identity transformation or a translation [see the solution to Problem 42(a), and in particular the remark following the solution]. Thus the "first" point A_{12} is obtained from A by the

sum of two translations (one or even both of them may be "translations through zero distance"—that is, the identity transformation); the "second" point (which we shall call A'_{12}) is obtained from A by the sum of the same two translations taken in the opposite order. The assertion of the problem follows from this [compare the solution to Problem 14(a)].

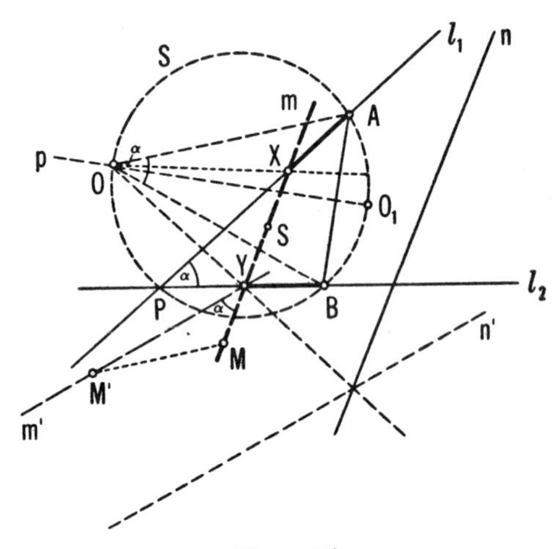

Figure 107a

45. *First solution* (based on Theorem 1, page 51). Suppose first that the lines l_1 and l_2 are not parallel (Figure 107a). Assume that the problem has been solved. By Theorem 1 the segment AX can be taken by a rotation into the congruent segment BY, so that A is taken into B and X into Y (since l_1 and l_2 are not parallel, AX cannot be taken into BY by a translation). The angle of rotation α is equal to the angle between l_1 and l_2; therefore the center of rotation O can be found as the point of intersection of the perpendicular bisector p of the segment AB with the circular arc constructed on AB and subtending an angle α (this arc lies on the circle S circumscribed about triangle ABP, where P is the point of intersection of l_1 and l_2).† Let this rotation take the desired line m into a line m', also passing through Y. We shall now consider Problems (a), (b), (c), and (d) separately.

† The circle S and the perpendicular bisector p intersect in two points O and O_1; they correspond to the cases when X and Y are situated on the same, or on opposite sides of the line through AB.

(a) Rotate the line n through an angle α about the center O that was found above, and let n' be the line thus obtained. The line OY will bisect the angles between m and m', and between n and n'; hence Y can be found as the point of intersection of l_2 with the line joining O to the point of intersection of n and n'. The problem can have two solutions (see the note†).

(b) m' passes through the point M' that is the image of M under a rotation through an angle α about the point O; the angle between m and m' is equal to α. Therefore Y can be found as the point of intersection of the line l_2 with the circular arc on MM' that subtends the angle α. The problem can have two solutions.

(c) In the isosceles triangle OXY we know the vertex angle α and the base $XY = a$; this enables us to find the distance OX from O to the unknown point X. The problem can have up to four solutions.

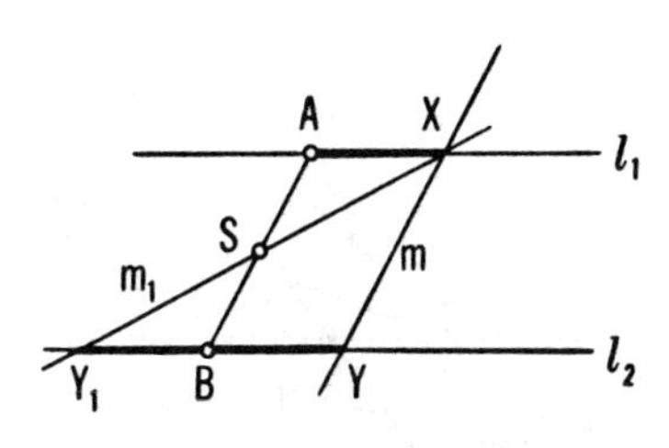

Figure 107b

(d) Let S be the midpoint of XY. Since the angles of the isosceles triangle OXY are known, we also know the ratio

$$\frac{OS}{OX} = k \qquad \text{and the angle} \qquad XOS = \frac{1}{2}\alpha.$$

Therefore the point S is obtained from X by a known spiral similarity (see Vol. 2, Chapter 1, Section 2).† The point S is found as the intersection of the line r and the line l_1' obtained from l_1 by this spiral similarity. The desired line m is perpendicular to OS. The problem has, in general, two solutions; if l_1' coincides with r then the solution is undetermined.

If $l_1 \parallel l_2$ then the desired line m either passes through the midpoint S of the segment AB or is parallel to AB (Figure 107b). In these cases the

† Here the second solution is preferable, as it does not use material from Vol. 2.

problem becomes much simpler. We shall merely indicate the number of solutions:

(a) One solution if n is not parallel to l_1 or to AB; no solutions if $n \parallel l_1 \parallel l_2$; infinitely many solutions if $n \parallel AB$.

(b) Two solutions if M does not lie on the line AB or on the line l_0 midway between l_1 and l_2 and parallel to them; one solution if M lies on AB or on l_0 but does not coincide with S; infinitely many solutions if M coincides with S.

(c) Two solutions if $a \neq AB$, and $a > d$ (where d is the distance between l_1 and l_2); one solution if $a = d$ but $AB \neq d$; no solutions if $a < d$; infinitely many solutions if $a = AB$ ($\geq d$).

(d) One solution if r is not parallel to $l_1 \parallel l_2$ and does not pass through S; no solutions if $r \parallel l_1$ but does not pass through S; infinitely many solutions if r passes through S.

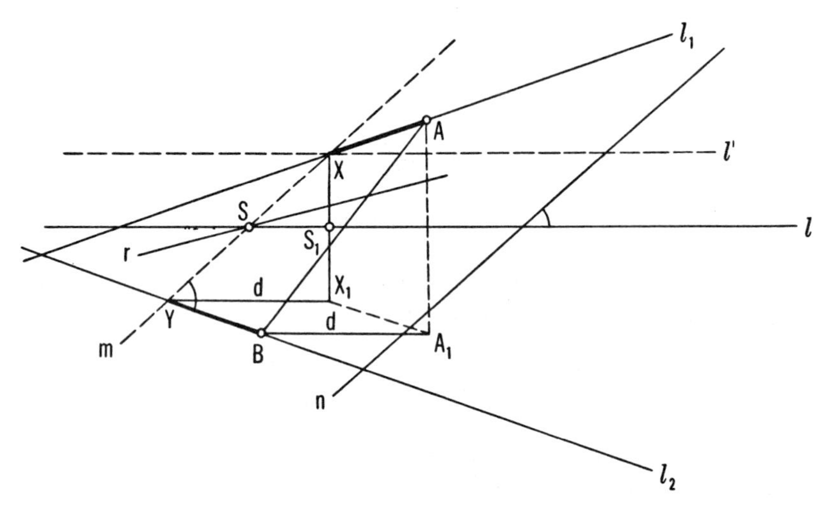

Figure 108

Second solution of parts (a), (c), (d) (based on Theorem 2, page 64). By Theorem 2 the segment AX can be taken by a glide reflection (or by an ordinary reflection in a line, which may be regarded as a special case of a glide reflection) into the congruent segment BY so that A goes into B and X into Y. Also, the axis l of the glide reflection is parallel to the bisector of the angle between l_1 and l_2 and passes through the midpoint of segment AB;† the distance d of the translation is equal to A_1B where A_1 is the image of A in l (Figure 108). Also, let X_1 be the image of X in l; in this case

$$X_1Y \parallel l \quad \text{and} \quad X_1Y = d.$$

We now consider the three cases (a), (c) and (d) separately.

(a) In triangle XX_1Y the side $X_1Y = d$ is known, as is $\measuredangle XYX_1$ (it is equal to the angle between m and l); hence the length of side XX_1 can be found. Now X can be found as the point of intersection of the line l_1 and the line l', parallel to l at a distance of $\frac{1}{2}XX_1$. In the general case, when l_1 is not parallel to l_2, the problem has two solutions.

(c) In triangle XX_1Y the hypotenuse $XY = a$ and the side $X_1Y = d$ are known; hence the other side XX_1 can be found. The remainder of the construction is similar to that in part (a); in general the problem has two solutions.

(d) The midpoint S of the segment XY must lie on the midline l of triangle XX_1Y. Therefore S is the point of intersection of l and r. X can now be found as the intersection of l_1 with the perpendicular p to l at the point S_1 (where $SS_1 = \frac{1}{2}d$). In general the problem has two solutions.

46. Suppose that the lines l_1, l_2, and l_3 are not all parallel to each other for example l_3 is not parallel to l_1 or to l_2. Assume that the problem has been solved (Figure 109). By Theorem 1 there is a rotation carrying AX into CZ and there is a rotation carrying BY into CZ; the angles of rotation α_1 and α_2 are equal respectively to the angles between l_1 and l_3, and between l_2 and l_3. The centers of rotation O_1 and O_2 are found just as in the first solution to Problem 45(a) – (d). From the isosceles triangles O_1XZ and O_2YZ with angles at O_1 and O_2 equal respectively to α_1 and α_2, one can find

$$\measuredangle O_1ZX = 90° - \tfrac{1}{2}\alpha_1, \qquad \measuredangle O_2ZY = 90° - \tfrac{1}{2}\alpha_2.$$

† Since there are two angle bisectors of the angles formed by l_1 and l_2, the glide reflection carrying AX into BY can be chosen in two different ways (corresponding to the cases when X and Y are situated on the same, or on opposite sides of the line AB). If $l_1 \parallel l_2$ then the axis of one of these glide reflections is parallel to l_1 and l_2 while the other axis is perpendicular to them; this explains the special role played by the case when l_1 and l_2 are parallel in the solution of parts (a), (c), (d).

From this it follows that

$$\measuredangle O_1ZO_2 = \tfrac{1}{2}(\alpha_1 \pm \alpha_2),$$

and, therefore, Z can be found as the point of intersection of l_3 with the arc of a circle constructed on the segment O_1O_2 and subtending the known angle $\tfrac{1}{2}(\alpha_1 + \alpha_2)$ or $\tfrac{1}{2}(\alpha_1 - \alpha_2)$.

Each of the angles α_1 and α_2, and each of the centers of rotation O_1 and O_2, can be determined in two different ways (compare the solution of the preceding problem). Hence there are at most 16 solutions to the problem.

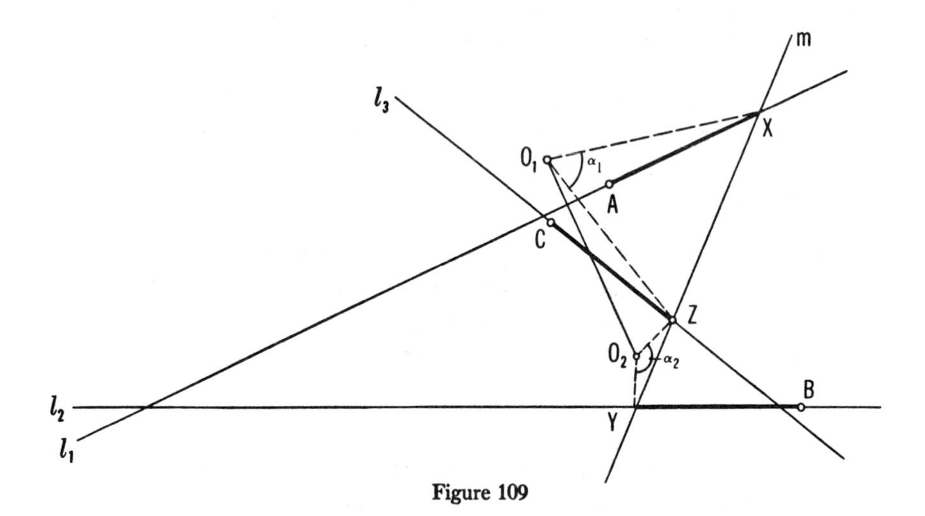

Figure 109

47. Assume that the problem has been solved (Figure 110). By Theorem 1 there is a rotation carrying BP into CQ; the angle of rotation α is equal to the angle between AB and AC, and the center of rotation O is found just as in the first solution to Problem 45(a) — (d). Since in the isosceles triangle OPQ we know the angle α at the vertex O, we also know the ratio

$$\frac{OP}{PQ} = k.$$

But by the conditions of the problem, $PQ = BP$; therefore

$$\frac{OP}{BP} = k,$$

which enables us to find P as the point of intersection of side AB with the circle that is the locus of points the ratio of whose distances to O

and B is equal to k. This geometric locus is a circle, as can be seen, for example, from the fact that the bisectors of the interior and exterior angles of ΔOPB from P (see Figure 111, where P is any point for which $OP/BP = k$) intersect the base OB in constant (independent of P) points M and N determined by the conditions

$$\frac{OM}{MB} = \frac{ON}{BN} = k = \frac{OP}{BP}$$

Since the two bisectors are perpendicular to each other, P belongs to the circle with diameter MN.[T]

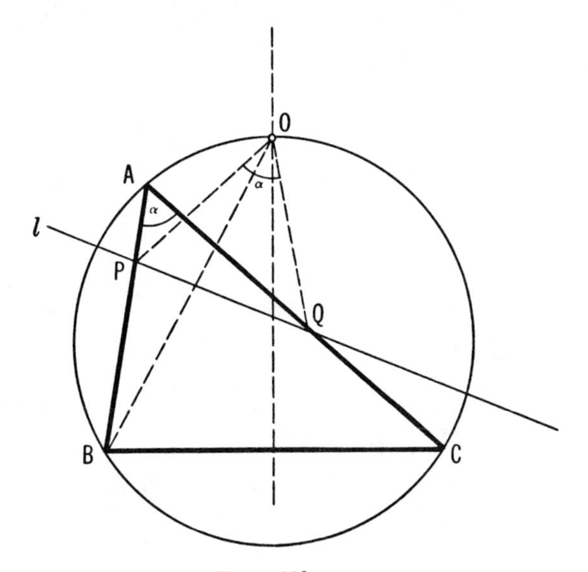

Figure 110

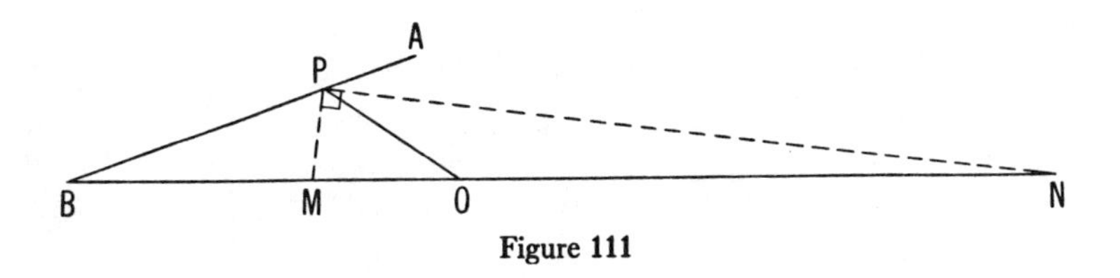

Figure 111

[T] See also page 14, Locus 11, of *College Geometry* by Nathan Altschiller-Court, Johnson Publishing Co., 1925, Richmond.